Raymond Merrill Smullyan
May 25, 1919 (age 94)

THE GÖDELIAN
PUZZLE BOOK

P. N. Ing 03/11/2014

THE GÖDELIAN PUZZLE BOOK

Puzzles, Paradoxes and Proofs

Raymond M. Smullyan

DOVER PUBLICATIONS, INC.
Mineola, New York

Bibliographical Note

The Godelian Puzzle Book is a new work, first published by Dover
Publications, Inc., in 2013.

International Standard Book Number
ISBN-13: 978-0-486-49705-1
ISBN-10: 0-486-49705-4

Manufactured in the United States by Courier Corporation
49705401   2013
www.doverpublications.com

PREFACE

This book, like hardly any others, explains the pioneering discoveries of the amazing logician Kurt Gödel through recreational logic puzzles. Its title is based on the fact that almost all the puzzles of the book are centered around Gödel's celebrated result.

In the first quarter of the twentieth century there were some mathematical systems in existence that were so comprehensive that it was generally assumed that every mathematical statement could either be proved or disproved within the system. In 1931 Gödel astounded the entire mathematical world by showing that this was not the case [Gödel, 1931]: For each of the mathematical systems in question, there must always be mathematical statements that can be neither proved nor disproved within the system. Indeed, Gödel provided an actual recipe for exhibiting in each such system, a sentence which must be true, but not provable, in the system. This famous result is known as *Gödel's Theorem*.

The essential idea behind Gödel's proof is this:

Gödel assigned to each mathematical sentence of the system a number, now known as the *Gödel number* of the sentence. He then constructed a most ingenious sentence S that asserted that a certain number n was the Gödel number of a sentence that was not provable in the system. Thus this sentence S was true if and only if n was the Gödel number of an unprovable sentence. But the amazing thing is that n was the Gödel number of the very sentence S! Thus S asserted that its own Gödel number was the Gödel number of an unprovable sentence. Thus in effect, S was a self-referential sentence that asserted its own non-provability. This meant that either S was true and not provable or S was false but provable. The latter alternative seemed completely out of the question since it was obvious from the nature of the system in question that only true sentences could be proved in the system. Thus Gödel's sentence S was true but not provable in the system. Its truth was known only by going outside the system and noting some of its properties.

How did Gödel manage to construct such an ingenious sentence? It is the purpose of this book to explain how, in terms that are completely comprehensible to the general public—even those with no background at all in mathematical logic. I have written this book so that it should be perfectly comprehensible to any reasonably bright high-school student. This is actually the first of my popular logic puzzle books in which I give a *complete* proof of Gödel's theorem for one particularly important mathematical system, as well as provide a host of generalizations that have never been published before, and should therefore be of interest, not only to the general reader, but to the logical specialist as well. These generalizations can be found in Chapters XIII, XIV and XV

On the whole, I have written this book in a very informal style. After the chatty introductory Chapter I, which consists mainly of personal anecdotes and jokes, the remaining chapters of Part I consist of puzzles, paradoxes, the nature of infinity (which often seems more paradoxical than it really is) and some curious systems related to Gödel's theorem. Part II is the real heart of this book, and could be read quite independently of Part I. The first three of its chapters contain my generalized Gödel theorems, which are unusual in that they do not involve the usual machinery of symbolic logic! I have deferred symbolic logic—the logical connectives and quantifiers—to the last three chapters, which begin by explaining its basics and what is known as *first-order arithmetic* followed by a presentation of the famous axiom system known as *Peano Arithmetic*. And I give a complete proof there of Gödel's celebrated result that there are sentences of Peano Arithmetic that cannot be proved or disproved within that axiom system.

Gödel's discoveries have led to an even more important result: Is there any purely mechanical method of determining which mathematical statements are true and which are false? This brings us to the subject of *decision theory*, better known as *recursion theory*, which today plays such a vital role in computer science. Chapter XVI of this book explains some basics of this important field. It turns out that in fact there is no purely mechanical method of deciding which mathematical statements are true and which are not! No computer can possibly settle all mathematical questions. It seems that brains and ingenuity are, and always will be, required. In the witty words of the mathematician Paul Rosenbloom, this means that "Man can never eliminate the necessity of using his own intelligence, regardless of how cleverly he tries!"

I would like to thank Dr. Sue Toledo for her very helpful editing work on this book.

TABLE OF CONTENTS

Part I—Puzzles, Paradoxes, Infinity and other Curiosities

Part II—Provability, Truth and the Undecidable

vii

PART I

PUZZLES, PARADOXES, INFINITY AND OTHER CURIOSITIES

CHAPTER I

A CHATTY PERSONAL INTRODUCTION

Let me introduce myself by what might be termed a *meta-introduction*, by which I mean that I will tell you of three amusing introductions I have had in the past.

1. The first was by the logician Professor Melvin Fitting, formerly my student, whom I will say more about later on. I must first tell you of the background of this introduction. In my puzzle book "What is the Name of this Book?" I gave a proof that either Tweedledee or Tweedledum exists, but there is no way to tell which. Elsewhere I constructed a mathematical system in which there are two sentences such that one of them must be true but not provable in the system, but there is no way to know which one it is. [Later in this book, I will show you this system.] All this led Melvin to once introduce me at a math lecture by saying, "I now introduce Professor Smullyan, who will prove to you that either he doesn't exist, or you don't exist, but you won't know which!"

2. On another occasion, the person introducing me said at one point, "Professor Smullyan is unique." I was in a mischievous mood at the time, and I could not help interrupting him and saying, "I'm sorry to interrupt you Sir,

but I happen to be the only one in the entire universe who is not unique!"

3. This last introduction (perhaps my favorite) was by the philosopher and logician Nuel Belnap Jr., and could be applicable to anybody. He said, "I promised myself three things in this introduction: First, to be brief, second, not to be facetious, and third, not to refer to this introduction."

I particularly liked the last introduction because it involved self-reference, which is a major theme of this book.

I told you that I would tell you more about Melvin Fitting. He really has a great sense of humor. Once when he was visiting at my house, someone complained of the cold. Melvin then said, "Oh yes, as it says in the Bible, many are cold but few are frozen." Next morning I was driving Melvin through town, and at one point he asked me, "Why are all these signs advertising slow children?"

On another occasion, we were discussing the philosophy of solipsism (which is the belief "I am the only one who exists!"). Melvin said, "Of course I know that solipsism is the correct philosophy, but that's only one man's opinion." This reminds me of a letter a lady wrote to Bertrand Russel, in which she said, "Why are you surprised that I am a solipsist? Isn't everybody?"

I once attended a long and boring lecture on solipsism. At one point I rose and said, "At this point, I've become an anti-solipsist. I believe that everybody exists except me."

Do you have any rational evidence that you are now awake? Isn't it logically possible that you are now asleep and dreaming all this? Well, I once got into an argument with a philosopher about this. He tried to convince me that I had no rational evidence to justify believing that I was now awake. I insisted that I was perfectly justified in being certain that I was awake. We argued long and tenaciously, and I finally won the argument, and he conceded that I did have rational evidence that I was awake. At that point I woke up.

Coming back to Melvin Fitting, his daughter Miriam is really a chip off the old block. When she was only six years old, she and her father were having dinner at my house. At one point Melvin did not like the way Miriam was eating, and said, "That's no way to eat, Miriam!" She replied, "I'm not eating Miriam!" [Pretty clever for a six-year old, don't you think?]

One summer, Melvin, who was writing his doctoral thesis with me, was out of town. We corresponded a good deal, and I ended one of my letters saying, "And if you have any questions, don't hesitate to call me collect and reverse the charges." [Get it?]

I would like to tell you now of an amusing lecture I recently gave at a logic conference in which I was the keynote speaker. The title of my talk was "Coercive Logic and Other Matters." I began by saying, "Before I begin speaking, there is something I would like to say." This got a general laugh. I then explained that what I just said was not original, but was part of a manuscript of the late computer scientist Saul Gorn about sentences which somehow defeat themselves. He titled this collection "Saul Gorn's compendium of rarely used clichés." It contains such choice items as:

1. Half the lies they tell about me are true.
2. These days, every Tom Dick and Harry is named "John."
3. I am a firm believer in optimism, because without optimism, what is there?
4. I'm not leaving this party till I get home!

5. If Beethoven was alive today, he would turn over in his grave!

6. I'll see to it that your project deserves to be funded.

7. This book fills a long needed gap.

8. A monist is one who believes that anything less than everything is nothing.

9. A formalist is one who cannot understand a theory unless it is meaningless.

10. The reason that I don't believe in astrology is because I'm a Gemini.

The last one was mine. I used that line frequently in the days that I was a magician. In those days, people often asked me whether I had ever sawed a lady in half. I always replied that I have sawed dozens of ladies in half, and I'm learning the second half of the trick now.

Next, I told the logic group that I had prepared two different lectures for the evening, and I would like them to choose which of the two they would prefer. I then explained that one of the lectures was very impressive and the other was understandable. [This got a good laugh].

Next, I said that I would give a test to see if members of the audience could do simple propositional logic. I displayed two envelopes and explained that one of them contained a dime and the other one didn't. On the faces of the envelopes were written the following sentences:

1. The sentences on the two envelopes are both false.

2. The dime is in the other envelope.

I explained that each sentence is, of course, either true or false, and that if anyone could deduce from these sentences where the dime was, he could have the dime. But for the privilege of taking this test, I would charge a nickel. Would anyone volunteer to give me a nickel for the privilege of doing this? I got a volunteer. I then told him,

"You are not allowed to just *guess* where the dime is; you must give a valid *proof* before the envelope is opened." He agreed. I said, "Very well. Where is the dime, and what is your proof?" He replied, "If the first sentence, the sentence on Envelope 1, were true, then what it says would be the case, which would mean that both sentences are false, hence the first sentence would be false, which is a clear contradiction. Therefore the first sentence can't be true; it must be false. Thus it is false that both sentences are false, hence at least one must be true, and since it is not the first, it must be the second, and so the dime must be in the other envelope, as the sentence says."

"That sounds like good reasoning," I said. "Open Envelope 1." He did so, and sure enough there was the dime.

After congratulating him, I said that the next test would be a little bit more difficult. Again I showed two envelopes with messages written on them, and I explained that one of them contained a dollar bill and the other was empty. The purpose now was to determine from the messages which envelope contained the bill. Here are the messages:

1. Of the two sentences, at least one is false.
2. The bill is in this envelope.

I then explained that if the one taking this test could correctly prove where the bill was, he or she could keep it, but for the privilege of taking this test, I would charge 25¢. After some thoughts, one man volunteered. I then asked him where the bill was, and to prove that he was right. He said, "If Sentence 1 were false, it would be true that at least one was false, and you would have a contradiction. Therefore Sentence 1 must be true, hence at least one of the sentences is false, as Sentence 1 correctly says. Therefore Sentence 2 is false, and so the bill is really in Envelope 1." I said, "Very well, open Envelope 1." He did so, and it was empty! He then opened Envelope 2, and there was the bill!

At this point, he, and other members of the audience looked puzzled. I then asked, "How come the bill was in Envelope 2 instead of Envelope 1?" One member of the audience yelled, "Because you obviously were lying!" I assured the audience that at no time did I lie, and indeed I never did! So given the fact that I did not lie, what is the explanation?

Problem 1. What is the explanation of why the bill was in Envelope 2, despite the volunteer's purported proof that the bill was in Envelope 1? What was wrong with the proof he gave? [Answers to problems are given at the end of chapters. Realize, though, that sometimes there are more ways than one to arrive at the solution to a given problem.]

At this point, the volunteer owed me 25¢. I then told the audience that I felt a little bit guilty about having won a quarter by such a trick. And so I said to the volunteer, "I want to give you a chance to win your money back, so I'll play you for double or nothing." [This got a general laugh.] "In fact," I continued, "I'll be even more generous!" I then handed him two $10 bills and told him that he could have his quarter back and even keep some of the money I just gave him, but he would have to agree to something first. I told him I was about to make a statement. If he wanted the deal I was proposing, he had to promise to give me back one of the bills if the statement was false. But if the statement turned out to be true, then he must keep both bills. "That's a pretty good deal, isn't it?" I asked. "You are bound to get at least $10, and possibly $20!" He agreed. I then made a statement such that in order for him to keep to the agreement, the only way was to pay me $1000!

Problem 2. What statement would accomplish this?

At this point the poor fellow owed me a thousand dollars. Later I will tell you how I gave him a chance (sic!) to regain his thousand dollars, but first I wish to tell you of a related incident (which I also told the audience): Many years ago, when I was a graduate student at Princeton, I would frequently visit New York City. On one of my visits I met a very charming lady musician. On my first date with her, I asked her to do me a favor. I told her that I would make a statement in a moment, and I asked her whether she would give me her autograph if the statement turned out to be true. She replied, "I don't see why not." And I said that if the statement was false, she should not give me her autograph. She agreed. I then made a statement such that in order for her to keep her word, she had to give me, not her autograph, but a kiss!

Problem 3. What statement would work?

Now, the statement I gave in the solution to the last problem had to be false, and she had to give me a kiss. However, there is another statement I could have made which would have had to be true, after which she would also have had to give me a kiss.

Problem 4. What statement could that be?

There is still another statement I could have made (a more interesting one, I believe) which could be either true or false, but in either case, she would have to give me a kiss. [There is no way of knowing whether the statement is true or false before the lady acts.]

Problem 5. What statement would accomplish this?

Anyway, whatever statement I would have made, it was a pretty sneaky way of winning a kiss, wasn't it? Well, what happened next was even more interesting. Instead of collecting the kiss, I suggested we play for double or nothing. She, being a good sport, agreed. And so she soon owed me two kisses, then with another logic trick four, then eight, then sixteen, then thirty-two, and things kept doubling and escalating and doubling and escalating and before I knew it, we were married! And I was married to Blanche, the charming lady musician, for over 48 years.

Once at breakfast I had the following conversation with Blanche:

Ray	Is *NO* the correct answer to this question?
Blanche	To what question?
Ray	To the question I just asked. Is *NO* the correct answer to that question?
Blanche	No, of course not!
Ray	Aha, you answered *NO*, didn't you!
Blanche	Yes.
Ray	And did you answer correctly?
Blanche	Why, yes!
Ray	Then *NO is* the correct answer to the question.
Blanche	That's right.
Ray	Then when I asked you what the correct answer is, you should have answered *YES*, not *NO!*
Blanche	Oh yes, that's right! I should have answered *YES*.
Ray	No, you shouldn't! If you answered *YES*, you would be affirming that *NO* is the correct answer so why would you give the incorrect answer *YES*?
Blanche	You're confusing me!

Fortunately, Blanche did not divorce me for this!

It's sometimes annoying for a wife to have an overly rational husband, isn't it? The following dialogue from my book "This Book Needs No Title" well illustrates this:

Wife   Do you love me?

Husband Well of course! What a ridiculous question!

Wife   You *don't* love me!

Husband Now what kind of nonsense is this?

Wife   Because if you *really* loved me, you couldn't have done what you did!

Husband I have already explained it to you that the reason I did what I did was *not* that I don't love you, but because of such and such.

Wife   But this such and such is only a *rationalization*! You really did it because of so and so, and this so and so would never be if you really loved me.

Etc., etc.!

Next Day

Wife   Darling, do you love me?

Husband I'm not so sure!

Wife   What!

Husband I thought I did, but the argument you gave me yesterday proving that I don't is not too bad!

I already told you how on my first date with Blanche, I won a kiss using logic. Here is another way of winning a kiss: I say to a lady, "I'll bet you that I can kiss you without touching you." After giving a precise definition of kissing and of touching, she realizes that it is logically impossible, and takes the bet. I then tell her to close her eyes. She does so, I then give her a kiss and say, "I lose!"

This is reminiscent of the prank in which you go into a bar with a friend who orders a martini. You place a tumbler on the martini and say, "I'll bet you a quarter that I can drink the martini without removing the tumbler." He accepts the bet. You then remove the tumbler, drink the martini and give him a quarter!

This is reminiscent of the story of a programmer and an engineer sitting next to each other on an airplane. The following conversation ensued:

Programmer   Would you like to play a game?
Engineer   No, I want to sleep.
Programmer   It's a very interesting game!
Engineer   No, I want to sleep.
Programmer   I ask you a question. If you don't know the answer, you pay me five dollars. Then you ask me a question and if I don't know the answer, then I pay you five dollars.
Engineer   No, no, I want to sleep.
Programmer   I'll tell you what! If you don't know the answer to my question, you pay me five dollars, but if I don't know the answer to your question, I'll pay you fifty dollars!
Engineer   O.K. Here's a question. What goes up the hill with four legs and comes down with five legs?

The programmer then took out his portable computer and worked on the question for an hour, but got nowhere. And so he handed the engineer fifty dollars. The engineer said nothing, but put the fifty dollars in his pocket. The programmer, a bit miffed, said, "Well, what's the answer?" The engineer then handed him five dollars.

Coming back now to my lecture and the guy who owed me a thousand dollars, I said to him, "I really feel sorry for you, and so I will give you back your thousand dollars on condition that you answer a yes/no question truthfully for me." He agreed. I then asked him a question such that the only way he could keep his word was by paying me, not a thousand dollars, but a million dollars!

Problem 6. What question would work?

At this point, I said to him, "I am now in a very generous mood, and so I'll tell you what I'm going to do! I'll give you back your million dollars on condition that you give me the answer to another yes/no question, but this time you don't have to answer truthfully! Your answer can be either true or false; you have the option! There is obviously no way I can trick you now, right?" He agreed that it was obviously impossible for me to con him under the given conditions, and so he accepted. Ah, but there was a way I could con him! The next question I asked was such that he had to pay me, not a million dollars, but a billion dollars!

Problem 7. How in the world was that possible?

Next, I told him that I was very sorry that he owed me a billion dollars, and so I would give him a 50% chance to win it back again, but for this privilege I would charge a nickel extra. "Isn't it worth a nickel," I asked, "for a fifty percent chance of winning back a billion dollars?" He agreed. I then wrote something on a piece of paper, folded it and handed it to someone so I could not use any slight of hand. I then explained that I had written a description of an event which would or would not take place in the room sometime in the next fifteen minutes. His job was to predict whether or not it would take place. "Your chances of predicting correctly is fifty percent, isn't it?" He agreed that it was. I then handed him a pen and a blank piece of paper and told him that if he believed that the event would take place, he should write "yes," otherwise write "no." He then wrote something on the paper. I asked, "Have you written down your prediction?" He said he had. I said, "Then you have lost!"

Problem 8. What could I have written such that regardless of whether he wrote "yes" or "no", he was bound to lose?

At this point, he still owed me a billion dollars. I then said to him, "I'll tell you what. I'll trade you the whole billion dollars for one kiss from your lovely wife!" [This got a real good laugh!]

I am incorrigible, you say? I certainly am! Indeed my epitaph will be:

IN LIFE HE WAS INCORRIGIBLE. IN DEATH HE'S EVEN WORSE!

Solutions to the Problems of Chapter I

1. In the first test I gave, I said that each sentence was either true or false. I never said that in the second test! If I don't say anything about the truth or falsity of the sentences involved, I can write anything I like and put the bill wherever I want! The fact is that in the last test, the sentence on Envelope 1 couldn't be either true or false. If it were false, you would have a logical contradiction. If it were true, you wouldn't have a *logical* contradiction; it would imply an empirically false fact—namely where the bill is. Thus the first sentence cannot be either true or false. [The second sentence, incidentally, is in fact true]. I use this puzzle as a dramatic illustration of Alfred Tarski's discovery that the very notion of truth is not well defined in various languages such as English. Later in this book I will give you a far more formal account of Tarski's theorem.

2. The statement I made was, "You will give me either one of the bills or a thousand dollars." If the statement was false, he would have to give me one of the bills, but doing so would make it *true* that he gives me either one of the bills or a thousand dollars, and we would have a contradiction. Hence my statement can't be false; it must be true. Therefore it is true that he must give me either one of the bills or a thousand dollars, but he can't give me one of his bills, because our agreement was that if my statement was true, he is to keep both bills! Therefore he has to give me a thousand dollars.

3. The statement I made was, "You will give me neither your autograph nor a kiss." If the statement were true, she would have to give me her autograph as agreed, but doing so she would make it false that she gives me neither her autograph nor a kiss, and we would have a contradiction.

Therefore the statement couldn't be true; it must be false. Since it was false that she was going to give me *neither*, then she had to give me *either*—either her autograph or a kiss. But she couldn't give me her autograph for a false statement, for that was the rule to which she had agreed! Hence she owed me a kiss!!

4. Here is a statement which has to be true and is such that she must give me a kiss. "Either you will not give me your autograph or you will give me a kiss."

I am asserting that one of the following two alternatives holds:

(1) You will not give me your autograph.

(2) You will give me a kiss.

If my assertion were false, making both alternatives false, then neither (1) nor (2) would hold, hence (1) would not hold, which means that she *would* give me her autograph, contrary to our agreement that she does not give me her autograph for a false statement. Therefore my statement cannot be false. Thus it is true that either (1) or (2) holds, but then she must give me her autograph, since the statement is true, which means that (1) cannot hold, and therefore (2) must hold. Thus she must give me a kiss (as well as her autograph).

5. A statement that works is, "You will give me either both an autograph and a kiss, or neither one." Thus I am asserting that one of the following alternatives holds:

(1) You will give me both.

(2) You will give me neither.

Suppose the statement is true. This means that one of the two alternatives really does hold, but it can't be (2), since she must give me her autograph for a true statement, hence it must be (1), and so she must give me both her kiss and her autograph.

Now suppose the statement is false. The only way it can be false is that she gives me one but not the other—either a kiss and no autograph, or an autograph and no kiss. The latter possibility is ruled out, because she cannot give me an autograph for a false statement. Therefore she must give me a kiss.

In summary, if the statement is true, she must give me both her autograph and a kiss, and if the statement is false, she must give me a kiss, but not her autograph. The interesting thing is that there is no way of knowing whether the statement is true or false, until the lady acts. It is actually up to her whether the statement is true or false! In either case, she must give me a kiss, but she has the option of giving me her autograph or not. If she does, that would make the statement true, and if she doesn't, that would make the statement false.

6. The question I asked was, "Will you either answer NO to this question or pay me a million dollars?" [Equivalently, I could have asked him the question, "Will you pay me a million dollars if you answer yes to this question?"]

I am asking whether one of the following two alternatives holds:

(1) You will answer NO
(2) You will pay me a million dollars.

If he answers NO, then he is claiming that neither alternative (1) or alternative (2) holds, whereas (1) did hold, so NO cannot be a correct answer. Hence to be truthful, he must answer YES. He therefore affirms that either (1) or (2) holds, but now (1) doesn't hold, and so it must be (2). Therefore he owed me a million dollars!

7. I said that he could answer me either truthfully or falsely. I never said that he could answer me paradoxically! Well, one can design a question such that unless he pays me a billion dollars, neither a YES nor a NO answer

could be either true or false, but paradoxical! Such a question is, "Is YES the correct answer to this question if and only if you pay me a billion dollars?" [In other words, is it the case that either yes is the correct answer to this question and you pay me a billion dollars, or *no* is the correct answer and you don't pay me a billion dollars]. If he doesn't pay me a billion dollars, then the question reduces to "Is *no* the correct answer to this question" and the answers *yes* and *no* are both neither true or false, but paradoxical, hence the only way he can avoid answering me paradoxically is by paying me a billion dollars.

8. What I wrote was "You will answer NO." If he wrote YES, then he is affirming that the event will take place, which it didn't, and if he writes NO, he is denying the event will take place, which it did. In either case he loses!

CHAPTER II

SOME CURIOUS ADVENTURES

Those of you who are familiar with some of my earlier puzzle books know about the place called the *Island of Knights and Knaves,* where knights always tell the truth and knaves always lie and every inhabitant is either a knight or a knave. Well, many years ago, long before I was married, I visited this strange place and had the following curious adventures, all of them leading to fascinating problems I had to solve. I will start with some simple ones.

Problem 1. On one of my visits I was introduced to three inhabitants A, B and C, and was told that at least one was a knight and at least one was a knave and that one of them had a prize that I could have, if I could determine which one had it. The three made the following statements:

A B doesn't have the prize.
B I don't have the prize.
C I have the prize.

Which one has the prize?

Problem 2. On my next visit to this island I met two natives named *Hal* and *Jal.* Hal uttered a statement of only three words, from which I could deduce that he and Jal were the same type (both knights or both knaves).

What statement could that have been?

Problem 3. On my next visit to this island I came across three natives A, B and C and was reliably informed that one of the three was a magician. They made the following statements:

A   B is not both a knave and a magician.
B   Either A is a knave or I am not a magician.
C   The magician is a knave.

Which one is the magician and what type is each?

Problem 4. A Court Case. I then witnessed a trial. A crime had been committed and three suspects, A, B, and C, were being tried. They made the following statements:

A   I am guilty.
B   I am the same type as at least one of the others.
C   We are all of the same type.

Which one is guilty?

Problem 5. On this particular island, each woman is either a constant liar or a constant truth teller. The men are as usual—knights and knaves. I was introduced to three married couples—the Arks, the Bogs, and the Cogs. One of the three couples is the king and queen of the island. I was reliably informed that in none of the couples are both of them liars. They all made the following statements:

Mr. Ark   I am not the king.
Mrs. Ark   The king was born in Italy.

Mr. Bog	Mr. Ark is not the king.
Mrs. Bog	The king was really born in Spain.
Mr. Cog	I am not the king.
Mrs. Cog	Mr. Bog is the king.

Which one is the king?

Problem 6. Another Court Case In this case, three couples—the Dags, the Eggs and the Fens were interrogated because it was known that one of the three men was a spy, but it was not known which one. A curious fact of this case is that in one of the couples, husband and wife were both truthful, in another, both were liars, and in another, one of the spouses was truthful and the other lied. They all made the following statements:

Mr. Dag	I am not the spy.
Mrs. Dag	Mr. Egg is the spy.
Mr. Egg	Mr. Dag is truthful.
Mrs. Egg	Mr. Fen is the spy.
Mr. Fen	I am not the spy.
Mrs. Fen	Mr. Dag is the spy.

Which one is the spy?

Problem 7. One day I saw an extremely beautiful lady on this island and was immediately smitten with her. I longed to know whether or not she was married, but I did not have the courage to ask her. The next day I came across her two brothers Alfred and Bradford. Alfred then made a statement. From this statement I could not tell whether or not the lady was married. Then to my surprise,

Bradford made the same statement, from which I could tell that she was not married.

What statement could that have been?

Problem 8. My next adventure on this island was quite harrowing! I got captured by a ferocious gang of brigands and was shown three natives A, B and C, and was told that one of them was the witch doctor. I was to point to one of them, and if I pointed to the witch doctor, I would get executed, but if I pointed to one who was not the witch doctor I could go free. The three made the following statements:

A   I am the witch doctor.
B   I am not the witch doctor.
C   At most one of us is a knight.

To which of the three should I point?

Problem 9. Actually the last adventure ended quite happily. I pointed to one of them and correctly explained why he couldn't be the witch doctor. The gang was quite pleased with my reasoning and became friendly. One of them said, "He seems like a nice guy; let us give him a reward!" They then showed me a picture of a very beautiful girl, and by Heavens, she was the very one with whom I was smitten! "She has seen you," one of them said, "and is quite fond of you. Tomorrow we will give you another test, and if you pass it, the lady will be yours."

True to their word, the next day I was shown five adjacent rooms and was told that the lady was in one of them. On the door of Rooms 1, 2, 3, 4, 5 were signs 1, 2, 3, 4, 5 respectively. From their signs I was to infer which room contained the lady, and which signs were true. If I succeeded, then the lady would be mine.

Here are the signs:

Sign 1.   The lady is not in Room 2.
Sign 2.   The lady is not in this room.
Sign 3.   The lady is not in Room 1.
Sign 4.   At least one of these five signs is false.
Sign 5.   Either this sign is false, or the sign on the room with the lady is true.

Which room contains the lady, and which of the signs are true?

Solutions to the Problems of Chapter II

1. If C has the prize then all three statements are true, which would mean that all three of the speakers are knights, contrary to what is given. If B has the prize then all three must be knaves, again contrary to what is given. Therefore it must be that A has the prize (and also A and B are knights and C is a knave).

2. What Hal said was, "Jal is truthful." If Hal is a knight, then Jal is truthful, as Hal said, hence also a knight. If Hal is a knave, then contrary to what he said, Jal is not truthful, hence also a knave.

3. From C's statement it follows that C cannot be the magician, because if he is a knight then the magician is really a knave and hence cannot be C. On the other hand if C is a knave, then contrary to his statement, the magician is not a knave but a knight, hence cannot be C who is a knave. Thus in either case, C is not the magician.

Next we will see that A must be a knight. Well, suppose he were a knave. Then his statement is false, which means that B must be both a knave and a magician. Since A is a knave (under our assumption) then it is true that *either* A is a knave *or* (anything else!). Thus it is true that *either* A is a knave *or* B is not the magician, but this is just what B said, and thus the knave B made a true statement which is not possible! Thus the assumption that A is a knave leads to an impossibility, hence A cannot be a knave. Thus A is a knight. Hence his statement is true, which means that B is not both a knave and a magician.

We now know the following:

(1) C is not the magician.

(2) A is a knight.

(3) B is not both a knave and a magician.

Next we will see that B cannot be the magician, for suppose he were. Then it is false that he is not the magician, and it is false that A is a knave (by (2)), hence both alternatives of B's statement are false. Hence B's statement must be false, which makes B a knave. Hence B is then both a knave and a magician, which is contrary to (3)! Thus it cannot be that B is the magician. Also C is not the magician, as we have seen. Thus it must be A who is the magician.

Also, since B is not the magician, what he said is true, hence B is a knight. As for C, what he said cannot be true, since the magician is really a knight (A), not a knave. Hence C is a knave.

In summary, A and B are both knights, C is a knave and the magician is A.

4. Clearly, if we can show that A is a knight, we will know that A is the guilty party, since that is what he claims. Now, if B is a knave, then A must be a knight, since B's telling a lie implies that he is the only knave. On the other hand, if B is a knight, then he really is of the same type as at least one of the others, as he said. Thus either A is a knight or C is a knight. If A is a knight, we are done. But if C is a knight, then all three really are of the same type, making A a knight again. Thus A is a knight, period! So A is unquestionably the guilty one.

5. If Mr. Cog is the king then both Mr. and Mrs. Cog are making false statements, contrary to what is given. Therefore Mr. Cog is not the king.

Now, Mr. Ark and Mr. Bog are in agreement, hence they are both knights or both knaves. If they were both knaves, their wives Mrs. Ark and Mrs. Bog would both be truthful which is impossible, since they can't both be right. Therefore Mr. Ark and Mr. Bog are both knights, hence their statements are both true, which means that

Mr. Ark is not the king. Therefore Mr. Bog is the king and Mrs. Bog is the queen.

6. If Mr. Dag is the spy, then the Dags are both knaves (since their statements are both false) and furthermore the Eggs must also both be knaves, which violates the given conditions. Therefore Mr. Dag is not the spy.

If Mr. Egg is the spy, then the Eggs and the Fens are both mixed couples, because Mr. Egg is then truthful, Mrs. Egg lied, Mr. Fen is truthful and Mrs. Egg lied. Again this cannot be, and so Mr. Egg is not the spy.

Thus the spy is Mr. Fen.

7. What Alfred said was, "Either at least one of us is a knave or she is not married."

If Alfred is a knave, then it would be true that at least one of them is a knave (namely Alfred), hence it would be true that either one of them is a knave or the lady is unmarried, but knaves don't make true statements, hence Alfred must be a knight. Hence it is true that *either* at least one is a knave *or* the lady is unmarried. If Bradford is a knight, then the lady must be unmarried (since it is then false that at least one is a knave), but if Bradford is a knave, then there it cannot be determined whether the lady is married or not. Also, there is no way of knowing whether Bradford is a knave or a knight. Thus until Bradford spoke, there was no way of knowing whether or not the lady was married. But after Bradford said the same thing, he agreed with Alfred who is a knight, hence Bradford must also be a knight, which then settles the case— the lady must be unmarried.

To show that several different approaches can be taken to solve most of these problems, here is another solution to Problem 7, which comes at the problem from a different direction:

What both Alfred and Bradford said was, "Either at least one of us is a knave or she is not married." Now let C be *either* Alfred or Bradford. If C is a knave, then at least one of the two brothers is a knave (namely C). Hence it would be true that either one of them is a knave or the lady is unmarried, which is what C said. But knaves don't make true statements, so C must be a knight. But since C was taken to be either Alfred or Bradford, both the brothers must be knights. Thus it must have been the second part of what each said that was true, and the lady had to be unmarried. Note that when only Alfred had spoken, all that could be deduced was that Alfred was a knight and either Bradford was a knave or the lady was unmarried.

8. I reasoned as follows: C is either a knight or a knave. Suppose he is a knight. Then what he said is true, hence A and B must both be knaves, and since B is then a knave, his statement is false, which means that he is the witch doctor. Thus if C is a knight then B is the witch doctor.

Now suppose C is a knave. Then contrary to what he said, there must be more than one knight, hence A and B are both knights. Since A is then a knight, his statement is true, which means that he is the witch doctor.

In neither case is C the witch doctor, and so I pointed to C.

9. Let us first look at Sign 4. If it were false, then it would be true that at least one of the signs is false (namely Sign 4), which would make Sign 4 true, and we would have a contradiction. Hence Sign 4 cannot be false; it must be true. Since it is true, then like it correctly says, at least one of the signs really is false.

Next let us consider Sign 5. If it were false, then both its claims would have to be false, and the first claim is that sign 5 is false, which would make Sign 5 true, and we would again have a contradiction. Since Sign 5 cannot be false, so it must be true. Since it is true, then, as it correctly says, either it is false or the sign on the room with the lady is true, but the first alternative is out, since the sign is not false, and so it must be the case that the sign on the room with the lady is true. We now know four things:

(1) Sign 4 is true.

(2) Sign 5 is true.

(3) The sign on the room with the lady is true.

(4) At least one of the five signs is false.

From (3) it follows that Sign 2 must be true, because if it were false, then, contrary to what the false sign says, the lady would be in Room 2, hence the sign on Room 2, which is the room with the lady, would be false, which by (3) is not the case. Therefore Sign 2 is true, and as it says, the lady is not in room 2, which makes Sign 1 also true. Thus Signs 1, 2, 4, and 5 are all true, and since at least one of the signs is false, it must be Sign 3. Hence, contrary to what Sign 3 says, the lady is really in Room 1. This solves everything.

Epilogue. And so I won the lady, but despite her beauty, I soon found that she never told the truth, and so we soon broke up, which turned out to be a good thing, since years later I met the very lovely pianist Blanche, with whom I was married for 48 years.

CHAPTER III

THE STRANGE ISLAND OF MUSICA

In some far off ocean there is a very, very, VERY strange island named MUSICA in which every inhabitant is either a pianist or a violinist, but no inhabitant is both. The amazingly curious thing about this island is that every lady pianist always tells the truth and every lady violinist always lies, whereas men are the opposite—the male pianists lie and the male violinists tell the truth! [How do you like this setup?]

Problem 1. In one of my visits to this weird place, I came across two inhabitants L (a lady) and M (a man) who made the following statements:

L   We are both pianists.
M   That is true.

Is the lady a pianist or a violinist? And what about the man?

Problem 2. In my next visit there, I met another lady and man whom we will respectively call L and M who made the following statements:

29

L   We both play the same instrument.
M   She is a pianist.

What instrument does each play?

Problem 3. Here is a rather easy puzzle about the strange island of Musica: On one of my visits there, I went to the home of a Mr. and Mrs. Smith. They owned a nice looking piano that I saw in their living room and I asked the wife whether it was a Steinway. She replied, "I am a violinist and our piano is not a Steinway."

Was the piano a Steinway or not?

Problem 4. On another of my journeys to Musica, I visited the home of a married couple. I sat on the sofa between the husband and wife. At one point the one on my left said, "My spouse is truthful." The other one then said, "My spouse is a pianist."

Who sat on my left, the wife or the husband?

Problem 5. A MUSICAL QUATRO: Suppose you receive an e-mail statement from a resident of the Island of Musica. Here are 4 questions:

(1) What statement could convince you that the sender must be a lady violinist?

(2) What statement could convince you that the sender was a lady pianist?

(3) What statement could convince you that the sender was a lady, in spite of your having no way of knowing what instrument she plays?

(4) What statement could convince you that the sender was a pianist, in spite of your having no way of knowing whether the sender was male or female?

Problem 6. Three more questions about Musica.

(1) What statement could be made by any inhabitant of Musica other than a lady violinist?

(2) What statement could be made by either a female pianist or a male pianist or a female violinist or a male violinist? Any member of the island of Musica could make that statement!

(3) I got an e-mail from a member of Musica from which I could deduce that the sender must be either a female pianist or a male violinist, but I had no way to tell which! Moreover the message had only two words! Can you supply such a statement?

Solutions to the Problems of Chapter III

1. Since the two agree, they are either both telling the truth or both lying. If they are both telling the truth, then they are both pianists, as both would be truthfully saying. But male pianists don't tell the truth, hence this possibility is out. And so both are lying, which means that L is a violinist and M is a pianist. [Quite a pair!]

2. From just L's statement alone, it follows that M must be a pianist. For suppose that L is a pianist. Then she is truthful, and hence he is a pianist like her. On the other hand, suppose she is a violinist. Then she lied, and so they play different instruments, which means that he is again a pianist. And so in either case he is a pianist. Since he is a male pianist, he lied, and so the two actually play different instruments, which means that she is a violinist—and also lied. [Again they both lied!]

3. She couldn't be a pianist, because if she were, she would be truthful and never would have said what she said, and so she is a violinist. Now if it were true that the piano was not a Steinway, then what she said would be true said—namely that she is a violinist and the piano is not a Steinway. But lady violinists don't make true statements. Therefore the piano must be a Steinway.

4. Let A be the one to my left and B the one to my right. Since A said that B is truthful, then the two are either both truthful or both lying (because if A is truthful then B is also truthful, as A truthfully said, and if A is lying, then B lies, contrary to what A falsely said). First consider the case that both are truthful. Then, as B truthfully said, A is a pianist. Thus A is a truthful pianist, and hence a lady. Now consider the case that both lied. Since B lied, then A must really be a violinist, and being a lying violinist, is again a lady. Thus in either case A is a lady, and so it is

the wife who sat on my left. There is no way of knowing whether she is a pianist or a violinist.

5. Here are the answers to the Musica Quatro:
(1) I am a male pianist.
(2) I am not a male violinist.
(3) I am a pianist.
(4) I am female.

Explanations (In case you need them):

(1) No truthful person would claim to be a male pianist (who lies), hence the statement was false, which means that the sender was not really a male pianist, but being a liar, must therefore be a female violinist.

(2) If the statement were false, that would mean that the person WAS a male violinist, but male violinists don't make false statements, hence the statement must be true. Thus it is true that the person is not a male violinist, but being truthful must therefore be a female pianist.

(3) A lady pianist could truthfully say that, and a lady violinist could lie and say that, but a male violinist, being truthful, would never say that. And since a male pianist lies, he would never make the true statement that he is a pianist. Thus any lady could say that, but no man could.

(4) A female pianist could truthfully say that and a male pianist could falsely say that, but a female violinist could not make the true statement that she is female, and a male violinist would never lie and claim to be female. Thus any pianist could say that, but no violinist could.

There is a pretty symmetry between (3) and (4): Only a female could claim to be a pianist, and only a pianist could claim to be female.

6. Here are the answers to the other three:
(1) I am not a male pianist.
(2) I am truthful.

(3) I exist. [that statement is obviously true, and only female pianists and male violinists make true statements.]

Epilogue. Line (3) just above reminds me of a joke about the philosopher Rene Descarte who begins his whole system with the famous words, I THINK, THEREFORE I EXIST. The story is told that Descarte was at a bar and the bartender asked him, "Monsieur Descarte, would you like a cocktail? Descarte replied, "I think not" and disappeared.

CHAPTER IV
FOUR METAPUZZLES

A metapuzzle is the type of puzzle that can be solved only on the basis of knowing that others could or could not solve it. I will start with a simple one.

Problem 1. A logician once visited the Island of Knights and Knaves and met two inhabitants named Arthur and Bernard and he asked Arthur, "Are both of you knights?" Arthur answered (he either said *yes* or *no*) and the logician then knew what type each one was.

What type is each one?

Problem 2. The logician then met three natives named *Archie, Barchie* and *Corey* and was reliably informed that one of them was the medicine man, and also that at least two of them were knaves. Archie and Barchie made the following statements:

Archie   I am the medicine man.
Barchie   I am not the medicine man.

The logician then asked Corey, "Is Archie really the medicine man?" Corey answered (*yes* or *no*) and the logician then knew which one was the medicine man.

Which one was it?

Problem 3. Next, the logician met two natives named *Hal* and *Jal*. He asked Hal, "Are both of you knights?" Hal answered (*yes* or *no*), but the logician couldn't tell which type each one was. The next day another logician met Hal and Jal and asked Hal, "Are both of you knaves?" Hal answered, and this logician also didn't know the type of each. The next day a third logician met the same pair and asked Hal, "Is it the case that you are a knight and Jal is a knave?" Jal answered but this logician also couldn't tell the type of each.

What type is each?

Problem 4. A logician once visited a very strange island in which not every inhabitant was either a knight or knave. Those who were neither one were called outcasts. An outcast would lie on some days and be truthful on others. On any day he would either lie the entire day or be truthful the entire day, but his behavior could change from day to day. On a given day, to say that a person is *currently* truthful is to say that he is either a knight (who is truthful on all days) or that he is an outcast and that today is one of his truthful days.

The logician was captured by a tribe of bandits. Curiously enough the chief bandit was a knight, and this fact was common knowledge. He had been interested in logic and loved to test his victims. And so he led the logician into a room with 9 natives and explained that only one of them was a knight—each one of the other eight was either a knave or an outcast. The logician's task was to determine which one was the knight. If he succeeded, he would be set free; otherwise, he would be executed. Here is what the nine men said.

Archie- The knight is either Cary, Elmak, Greg or myself.

Barab	I am an outcast.
Cary	Either Elmak is currently truthful or Greg is not.
Dreg	Archie lied.
Elmak	Barab and Dreg didn't both lie.
Frisch	Cary lied.
Greg	Archie is not the knight.
Hal	I am a knave and Ilak is an outcast.
Ilak	I am a knave and Frisch lied.

The logician thought about this for a while and finally said, "I don't have enough information to solve the problem. I need to know whether or not Hal is an outcast. Then I might be able to solve it." The chief bandit was fair enough to tell him whether or not Hal was an outcast, and the logician could then figure out which one was the knight.

Which one was the knight?

Solutions to the Problems of Chapter IV

1. Suppose Arthur answered *yes*. Then there are these possibilities: (1) Both are knights; (2) Arthur is a knave and Bernard is a knight; (3) Both are knaves. But the logician would have no way of knowing which of those three possibilities held.

On the other hand, suppose Arthur answered *no*. There is then only one possibility: If Arthur were a knave then it would be false that both are knights, hence Arthur's answer *no* would be correct, but knaves don't answer questions correctly. Hence Arthur cannot be a knave; he is a knight. Hence his *no* answer was correct, which means that they are not both knights, and so Bernard must be a knave.

Thus if Arthur answered *yes*, the logician would have no way of knowing the type of either Arthur or Bernard, whereas if Arthur answered *no*, then the logician would know that Arthur was a knight and Bernard was a knave. But we are given that the logician *did* know, hence Arthur must have answered *no* and Arthur is a knight and Bernard a knave.

2. Suppose Corey answered *yes*. Then he agrees with Archie, hence they are either both knights or both knaves. They can't both be knights, for it is given that at most one is a knight. Hence they are both knaves. Thus Archie is not the medicine man. It could be that Barchie is the medicine man and a knave, or it could be that Corey is the medicine man, and there is no way to tell which. Thus if Corey answered *yes*, the logician couldn't have known which one was the medicine man. But the logician *did* know, hence Corey must have answered *no*.

There is then only one possibility. Since Corey answered *no* he disagreed with Archie, hence the two are of different types—one of them is a knight and the other a knave. Then Barchie must be a knave, or there would be two knights. Thus Barchie is a knave, and contrary to what he said, he is the medicine man.

3. As the reader can verify, if Hal answered *no* to the first logician, then the only possibility is that Hal is a knight and Jal is a knave, and the logician would then have known that Hal is a knight and Jal a knave. But since the logician didn't know the type of each, then Hal must have answered *yes*, and all the logician could then know is that it is not the case that Hal is a knight and Jal is a knave. Thus Hal answered *yes* and it is not the case that Hal is a knight and Jal is a knave.

As for the encounter with the second logician, had Hal answered *yes*, the only possibility would be that Hal is a knave and Jal a knight (as the reader can verify), and the logician would then have known that Hal was a knave and Jal a knight, but since he didn't know, then Hal must have answered *no*, from which it follows that it cannot be that Hal is a knave and Jal is a knight.

We now see that it cannot be that Hal is a knight and Jal a knave, nor can it be that Hal is a knave and Jal is a knight. Therefore Hal and Jal are either both knights or both knaves.

As to the encounter with the third logician, if Hal had answered *no* then the only possibility would be that Hal and Jal are both knights, and the logician would then have known that they were both knights, but the logician didn't know, hence Hal must have answered *yes*, from which it follows that the two are not both knights. But since they are of the same type, they are both knaves.

4. If the logician had been told that Hal was an outcast, he couldn't have solved the problem, but since he did solve it, he must have been told that Hal wasn't an outcast.

And so we now know that he wasn't. Thus he is either a knight or a knave. He obviously is not a knight, because of what he said, and so he is a knave. Now, if Ilak was an outcast, Hal's statement would be true, which it isn't. Therefore Ilak is not an outcast. From his statement, he is obviously not a knight, hence he is a knave. Thus Hal and Ilak are both knaves. If Frisch lied, then Ilak's statement would be true, hence Frisch did not lie. Therefore Corey lied, as Frisch said. Since the statement by Corey is false, then neither of the two alternatives of the statement hold, which means that Elmak lied and Greg told the truth. Since Elmak lied, Barak and Dreg both lied. Since Greg told the truth, then Archie is not the knight. Since Dreg lied, then Archie told the truth. Thus Archie told the truth, but is not the knight. Then as Archie truthfully said, the knight is either Corey, Elmore or Greg. But Corey and Elmore both lied, as we have seen, hence Greg is the knight.

CHAPTER V
CERTIFIED KNIGHTS
AND KNAVES

Many of the problems of this chapter are related to Gödel's theorem, as the reader will later see.

I once visited an island of knights and knaves in which certain knights have been *proven* to be knights. They are called *certified* knights. Also some of the knaves have been proven to be knaves, and are called *certified* knaves. Those who are not certified are called *uncertified*.

Problem 1. I once came across a native of the island who said, "I am not a certified knight." Was he a knight or a knave? Was he certified?

Problem 2. On another occasion I came across a native who said, "I am an uncertified knight." Is this the same situation as the last problem? Is it now possible to tell whether or not he is a knight? Can one tell whether or not he is certified?

Problem 3. I once came across a native who made a statement from which I could deduce that he must be a uncertified knave. What statement would work?

Problem 4. I once came across a native who made a statement such that I could deduce that he must be certified (either a certified knight or a certified knave), but there was no possible way to tell whether he was a knight or a knave.

What statement would work?

Problem 5. Another time a native made a statement such that I could infer that he must be either a certified knight or an uncertified knave, but there was no way to tell which.

What statement would work?

Problem 6. On another particularly interesting occasion I came across two natives A and B each of whom made a statement such that I could infer that at least one of them must be an uncertified knight, but there was no way to tell which one it was. From neither statement alone could one have deduced this.

What two statements would work? [This problem is not easy!]

Problem 7. On another occasion I came across two natives A and B. I was curious to know of each one whether he was a knight or a knave and whether or not he was certified. They made the following statements:

A   B is an uncertified knight.
B   A is a certified knave.

I thought awhile and could solve part of the problem, but not all of it, because I had no way of knowing whether B was a knight or a knave. I later found out whether B was a knight or a knave, and then could solve the entire problem! What is the solution?

Problem 8. Now for my most glorious adventure! On this island, each woman is also a constant liar or a constant truth-teller. Also, women are classified as either *certified* or *uncertified* (liars or truth-tellers). Well, I was given a test to see whether I could reason logically. I was introduced to two married couples. The ladies were named *Beatrice* and *Cynthia*, and the husbands were named *Kenneth* and *Walter*, but I was not told who was married to whom. Each of the four made a statement and I was supposed to deduce who was married to whom, which ones were certified and which ones lied and which ones told the truth. Here is what they said:

Kenneth	My wife is untruthful.
Walter	Either my wife is untruthful or she and I are both certified truth-tellers.
Beatrice	My husband is truthful.
Cynthia	My husband is truthful and at least one of the two of us is uncertified.

What is the solution?

Epilogue. I passed the exam with flying colours, and the inhabitants were so delighted and impressed that they pooled together and gave me a cash prize, and the University of Knights and Knaves gave me an honorary degree, and best of all, I was kissed by all the lovely ladies of the island!!

[As you can see, I'm not very modest!]

Solutions to the Problems of Chapter V

1. If he were a knave, then it would be true that he is not a certified knight, but knaves don't make true statements. Therefore he can't be a knave; he must be a knight. Since he is a knight, his statement must be true, which means that he is an uncertified knight, as he said.

2. This is very different! If he were a knave, it would be *false* that he is an uncertified knight, and knaves are capable of making false statements, hence he could be a knave—certified or otherwise. He also could be a knight, but only if he is an uncertified knight. Thus, unlike Problem 1, he could be any one of these things: (1) a certified knave; (2) an uncertified knave; (3) an uncertified knight.

3. A statement that works is, "I am a certified knave." A knight couldn't say that, hence the speaker was a knave. It then follows that contrary to what he said, he is not a certified knave, but since he is a knave, he must be an uncertified knave.

4. A statement that works is "I am either a certified knight or an uncertified knave."

Suppose he is a knight. Then his statement is true, hence he really is either a certified knight or an uncertified knave, but the latter alternative cannot be (he cannot be an uncertified knave, since he is not a knave), hence he is a certified knight, hence he is certified.

On the other hand suppose he is a knave. Then his statement was false, which means that he is neither a certified knight, nor an uncertified knave. Thus he is not an uncertified knave, yet he is a knave, hence he must be a certified knave.

This proves that regardless of whether he is a knight or a knave, he must be certified. [There is no way of knowing whether he is a knight or a knave].

5. There are many statements that could work, but a particularly simple one is "I am certified." If he is a knight, then he really is certified, as he said, hence he is a certified knight. If he is a knave, then contrary to what he said, he is not certified, hence he is then an uncertified knave. Thus he is either a certified knight or an uncertified knave, and there is no way to tell which.

6. A and B made the following statement:

A   B is a certified knight.
B   A is not a certified knight.

Suppose A is a knight. Then B is a certified knight, as A truthfully said, hence B is a knight, hence A is not a certified knight, as B truthfully said, hence A is then an uncertified knight.

On the other hand, suppose A is a knave. Then what B said was true (A is obviously not a certified knight), hence B is a knight. But since A is a knave, his statement was false, which means that B is not a certified knight. Thus B is a knight but not a certified knight, hence he is an uncertified knight.

In summary, if A is a knight then he is an uncertified knight, but if he is a knave, then B is an uncertified knight. There is no way of knowing which of the two cases holds.

7. Before I found out who told the truth, I reasoned as follows: Suppose A is a knight. Then his statement is true, which means that B is a uncertified knight, hence a knight, hence his statement is true, which means that A is a certified knave, contrary to the assumption that A is a knight. Thus A cannot be a knight; he must be a knave. And so at this point I know that A was a knave, and therefore, contrary to what he said, B was not really

an uncertified knight. Next I considered the two possible cases of whether B was a knight or a knave.

Case 1: Suppose B is a knight. Then, since I already knew that B is not an uncertified knight, then B must be a certified knight. Also, if B is a knight, his statement must be true, which means that A is a certified knave. Thus if B is a knight, the whole problem is solved—B is a certified knight and A is a certified knave.

Case 2: But suppose B is a knave? Then his statement is false, which means that A is *not* a certified knave, but since he is a knave, he is therefore an uncertified knave. And so all I could conclude from this possibility is that A is an uncertified knave and B is a knave, but there was no way of knowing whether B was certified or not. Also I had no way of knowing whether B was a knight or a knave. So I was completely in the dark. Then, as I told you, I found out (by another means) whether B was a knight or a knave. Had I been told that B was a knave, I would have had no way of knowing whether or not B was certified, as I already explained. But, as I told you, I *did* solve the entire problem, and so I was told that B is a knight. Then I knew for sure that B is a certified knight and A is a certified knave.

8. It is obvious that Kenneth cannot be married to Beatrice, because if he was then he would have the following contradiction: If Kenneth is truthful then Beatrice lied, as Kenneth said. Then contrary to what she said, Kenneth is not truthful, which is a contradiction. On the other hand, if Kenneth lied, then contrary to what he said, Beatrice is truthful, hence, as she said, Kenneth is truthful, and we again have a contradiction. Therefore Kenneth is not married to Beatrice; he is married to Cynthia. Thus also, Walter is married to Beatrice.

Let us now consider Walter and Beatrice. Suppose Beatrice lied. Then contrary to what she said, Walter lied. This means that both alternatives of Walters statements are false, hence the first alternative is false, which means that Beatrice is truthful, and we thus have a contradiction. Therefore it cannot be that Beatrice lied; she told the truth, and so Walter is truthful, as Beatrice said. Now, the first of Walters alternatives doesn't hold, as we have seen, hence the second one does (since Walter's statement is true) and so both Walter and Beatrice are certified truth tellers.

Now let us consider the Cynthia Kenneth couple. If Kenneth lied, then Cynthia is truthful, and hence Kenneth is truthful, as Cynthia said. This is an obvious contradiction, and so Kenneth is truthful. Therefore Cynthia lied, as Kenneth said. If at least one of the two was uncertified, then Cynthia's statement would be true, which it isn't, and so neither one is uncertified—both are certified.

We thus see that Walter is married to Beatrice, Kenneth is married to Cynthia, all four of them are certified and Cynthia is the only one who lied.

This solves everything.

CHAPTER VI

PARADOXICAL?

Paradoxes are a lot of fun. But first for a little riddle:

Problem 1. Why is it logically impossible for there to be more than one doctor in the universe?

Here is a little paradox I recently thought of: Let us define a *charlatan* as one who is not what he pretends to be. Well, suppose someone pretends to be a charlatan, but isn't one really—he is only pretending to be one. But doesn't his false pretense make him a charlatan? Now that we know he is a charlatan, then he really is what he pretended to be, and so in what sense is he a charlatan?

I hope this will be known as the Smullyan Charlatan Paradox, and not as the Charlatan Smullyan Paradox!

A mathematician friend of mine told me that once on an automobile trip to Canada he came across a sign in the road which said, "Please ignore this sign."

I like a paradox invented by the literary agent Lisa Collier about the president of a company who offered a hundred dollar reward to any employee who could make a suggestion that would save money for the company. One employee suggested, "Eliminate the reward!"

Paradoxes go back to ancient times. One of the oldest paradoxes is of course Zeno's famous paradox, proving that motion is impossible. Suppose a body is to move from a point A to a point B. Starting from A, before it can get to point B it must get to a point A_1 midway between A and B. That is the first step. Before it can get from A_1 to B, it must first get to point A_2 midway between A_1 and B. That is the second step. The third step is to go from A_2 to a point A_3 midway between A_2 and B, and so forth. After any finite number of such steps it has not yet reached B. Thus no finite number of steps can suffice for the body to reach B, and so motion is impossible!

Problem 2. What is wrong with Zeno's arguments? If an argument leads to a false conclusion, there must be a first false step in the argument. What is the first false step in Zeno's first argument?

Zeno's argument reminds me a little of an argument I once heard that it is impossible for one to die: If he dies, at what instant does he die? Is the instant while he is alive or when he is dead? He can't die at the instant while he is still alive, or he would be alive and dead at the same time, and he can't die when he is dead, because he is already dead. Therefore he can't die at all!

Now let us go back to ancient paradoxes.

The Protagoras Paradox. There is the story of a law teacher named *Protagoras* and a talented youth who wanted to study with him but had no money to pay. Well, Protagoras made the following contract with the boy: Protagoras would teach him without pay, but in return, after the instruction was over and the boy could then practice law, he would pay Protagoras an agreed amount after winning his first case. Well, after the studies were over and the boy could now practice law, he didn't take any cases, so Protagoras sued him and the boy acted as his own lawyer. He argued as follows: Either I win this case or I lose this case. If I win it, it means I don't have to pay, since this case is what the issue is about. On the other hand, if

I lose this case, I won't yet have won my first case, hence I won't have to pay. In either case, I don't have to pay.

Protagoras then spoke and gave the following argument: He has it all wrong! If he loses this case, I will have won it, which means that he has to pay me. On the other hand, if he wins this case, he will have won his first case and therefore will have to pay me. In either case, he must pay me.

Problem 3. How should this be decided?

In the Plato dialogues there is the sophist Protagoras of whom Socrates said to a friend, "So great is his skill that he can prove any proposition, whether true or false!" Indeed, Protagoras once gave the following proof to Socrates that Socrates was the son of a dog!

Protagoras	Do you have a dog?
Socrates	Yes.
Protagoras	So it is your dog.
Socrates	Yes.
Protagoras	Does the dog have puppies?
Socrates	Yes, I saw him and the mother come together.
Protagoras	So he is a father.
Socrates	Yes.
Protagoras	And he is yours.
Socrates	Yes.
Protagoras	And so he is YOUR FATHER, and the puppies are your brothers and sisters.

On another occasion, Socrates was chiding Protagoras for taking money to teach students wisdom. Protagoras then explained to Socrates that at the end of the instruction, if the student feels like he hasn't learned enough, his money is refunded in full! When I read this, I couldn't help thinking of the following scenario:

A student studies with Protagoras and at the end of the instruction, he tells Protagoras that he hasn't learned enough and demands that he be given his money back. Protagoras asks the student whether he can give a good argument why Protagoras should give him his money back. The student then gives an excellent argument, upon which Protagoras says, "You see the dialectical skill that I have taught you!"

Next, another student, after his course with Protagoras, tells him that he hasn't learned enough, and demands that Protagoras give him his money back. Again, Protagoras asks him to give a good argument proving that Protagoras should give him back his money. The student then gives a very poor argument, upon which Protagoras says, "O.K., here's your money back!"

Russel's Paradox and Related Ones. At the turn of the century there appeared an extremely powerful mathematical system—Set Theory, by Gottlöb Frege. The only trouble with it was that it is inconsistent, as pointed out by Bertrand Russel:

One of Frege's axioms of his system was that given any property of sets, there exists the set of all things having the property. This seems perfectly reasonable on the surface, but as Russell pointed out, it leads to a contradiction: Call a set *ordinary* if it is not a member of itself. For example, the set of all points on a given line is not itself a point, hence this set is ordinary. Or again, the set of all people on this planet is not a person hence that set is ordinary. Virtually all sets that we come across are ordinary, and it is questionable whether there exist any sets that are not ordinary. At any rate, ordinary sets certainly exist, and according to Frege's axiom mentioned above, one can talk about the set of all ordinary sets—Call this set "S." Is S itself ordinary or not? Either way we have a contradiction: Suppose S is ordinary. Then, since S contains *all* ordinary sets, it must in particular contain the ordinary set S, and thus S is a member of S which means that S is not ordinary. So it is impossible that S is ordinary. Now suppose S is not ordinary. This is also impossible, since this then means by definition that S is a member of S, contrary to the fact that only ordinary sets are members of S. The upshot is that there cannot be such a thing as the set of all ordinary sets, and so Frege's axiom system is inconsistent.

Frege was quite heartbroken when he received Russel's proof of inconsistency, and felt that his whole life work was a failure. Actually, this was quite far from the truth! Only minor modifications of Frege's system were necessary to produce a powerful and quite comprehensive system of set theory which appears to be consistent. Such systems would probably never have originated without the pioneering work of Frege. He and Georg Cantor are rightfully regarded as the founding fathers of set theory.

The philosopher and logician Bertrand Russel gave the following popular characterization of the set theory paradox:

Consider a barber of a certain town who shaves all the inhabitants of the town who don't shave themselves, and only such inhabitants. In other words if an inhabitant doesn't shave himself, the barber shaves him, but the barber never shaves any inhabitant who shaves himself. The problem is this: Does the barber shave himself or doesn't he? If he does, then he is shaving someone who shaves himself, contrary to the given condition that he never does that. On the other hand, if he doesn't shave himself, then he is failing to shave someone who doesn't shave himself, and this also contradicts the given conditions. Hence it is impossible either way!

Problem 4. How do you get out of this one?

Remark. I must tell you of an amusing incident. Remember that in the statement of the Barber Paradox, the barber of a certain town shaves all and only those inhabitants *of the town* who don't shave themselves. Well, I once told this paradox to that clever lady musician with a good sense of humor already mentioned earlier. She replied, "Oh, he probably went to his brother's house in another town and shaved himself!"

I loved that!

Closely related to Russel's paradox is a paradox known as Grelling's paradox: Call an adjective *Autological* if it itself has the property that it denotes, and *Heterological* if it doesn't. For example the word "polysyllabic" is autological, because the word itself has more than one syllable, and is therefore polysyllabic. Thus the word "polysyllabic" is polysyllabic, hence is autological. On the other hand, the word "monosyllabic" is heterological, because the word has more than one syllable, hence is not monosyllabic. Again, the word "red" if printed in red would be autological, but if printed in any other color, would be heterological. Now, what about the word "heterological"? Is that word autological or heterological? Either way we get a contradiction. If it is heterological, then it has the very property that it describes—the property of being heterological, which makes it autological. On the other hand, if it is autological, then it applies to itself, which means that it is heterological. Thus it can't be either without being both, which is logically impossible.

A Logicians Escape. A logician inadvertently entered a land that was forbidden to outsiders. He was immediately captured and told the decree of the land, namely that any one who entered from outside had to make a statement. If the statement was true, he would be hanged, and if the statement was false, he would be drowned. Well, the logician cleverly made a statement such that it was impossible for the inhabitants to carry out their decree.

Problem 5. What statement would work?

Newcomb's Paradox. Consider a chest of two drawers. Either there is $100 in each drawer or there is $1000 in each drawer; you don't know which. Thus the chest contains either $200 or $2000. You have the option of choosing just the money in the bottom drawer or the money in both drawers. Which option would you pick? Which way would you get more money, by choosing just the bottom drawer or choosing both drawers? Surely you would opt to take the money from both drawers, since there is twice as much money in both than in just one. Now, is there any further information I could give you that would make you change your mind? Most people to whom I have posed this problem say that there is not. Well, what I didn't tell you is that there is somewhere a perfect predictor—a God, a computer, or some being that has complete knowledge of the future and knows what choice you will make. This being also has control of how much money goes into the drawers. If he predicts that you will choose both drawers, he will put $100 in each drawer, but if he predicts that you will choose just the bottom drawer, he will put $1000 in each drawer. Does this information change your mind? Well, there are two schools of thought about this. There are those (including myself) who would choose just the bottom drawer, on the grounds that if I choose both, I will get only $200, whereas if I choose just the bottom drawer, I will get $1000. The other school says, "Nonsense! The money is already there. There is twice as much money in both drawers than in the bottom drawer. Therefore you should take both drawers.

I was told about this paradox by Martin Gardner. I told him that of course I would choose just the bottom drawer. He replied, "Don't be too hasty, Raymond! Suppose the backs of the drawers were made of glass and friends of yours were behind the drawers and could see how much money was in each drawer. They want you to get as much money as possible. Wouldn't they hope that you would choose both drawers?" I didn't know how to answer that at that time, but I subsequently realized that the only way they could hope that I choose both drawers is that they either didn't know about the predictor, or knew about but didn't trust the predictor. If they knew about and trusted the predictor, they would know how I would choose, and couldn't rationally hope that I would choose otherwise.

In one book in which Martin discussed this paradox, he comes to the conclusion that it proves that it is logically impossible that there is a perfect predictor. I must respectfully disagree with Martin on that point. I believe that the predictor is not really essential to the problem, and I thought of a version of the paradox in which the predictor doesn't even appear!

Here is my version:

The setup conditions for the drawers is the same as before—either that there is $100 in each or $1000 in each. Now consider the following proposition:

Proposition A. Either (1) You will choose both drawers and there is $100 in each drawer, or (2) you will choose just the bottom drawer and there is $1000 in each drawer.

Note that in the above proposition, no mention is made of a predictor. The function of the predictor in the original version was to give plausibility to Proposition A.

Now, if Martin's claim is correct, does that mean that Proposition A is inconsistent? Is Proposition A consistent? First I will prove that the proposition is inconsistent, then I will prove that the proposition is consistent. That is my version of the paradox.

Proof that Proposition A is Inconsistent. It follows from Proposition A that if you choose just the bottom drawer you will get $1000, whereas if you choose both drawers you will get $200, therefore you will get more money by choosing just the bottom drawer. On the other hand, there is twice as much money in both drawers as in just the bottom drawer, and hence you will get more money by choosing both drawers than by choosing just the bottom drawer. This is a clear cut inconsistency!

Proof that Proposition A is Consistent. To show that a proposition is consistent, if suffices to exhibit a situation in which it is true. Well, it certainly is possible that you will choose both drawers and find $100 in each, which would validate the proposition (and it's also possible that you choose just the bottom drawer and that there is $1,000 in each drawer). And so the proposition is consistent after all.

This is my version of the paradox, and as I already said, no predictor is involved.

A Curious Paradox. Here is a curious one: We have two positive whole numbers x and y and we are given that one of them is twice the other. I will now prove to you the following obviously incompatible propositions:

Proposition 1. The excess of x over y, if x is greater than y, is greater than the excess of y over x if y is greater than x.

Proposition 2. The two excesses are the same.

Proof of Proposition 1. If x is greater than y, x is twice y, hence the excessive of x over y is y.

If y is greater than x, then x is 1/2y, hence the excess of y over x is 1/2y. Obviously y is greater than 1/2y. Thus proves Proposition 1.

Proof of Proposition 2. Let d be the positive difference between x and y. (d is x − y if x > y, and is y − x, if y > x). Well, if x is greater than y, the excess of x over y is d, and if y is greater than x, then the excess of y over x is also d! Therefore the possible excesses are the same!

Hm!

Solutions to the Problems of Chapter VI

1. Because if there was so much as two docs in the universe you would have a pair-a-docs.

2. This is about the only argument I know in which the first false step is the conclusion! Everything up to there is correct—it is indeed true that to get from A to B one must go through infinitely many steps of the sort that Zeno described, but so what? Zeno never proved that you cannot go through infinitely many steps in a finite length of time, and it cannot be proved, because it is false!

3. The cleverest solution I heard was from a lawyer: He suggested that the student should win the case, since he has not won any prior cases, and accordingly should not have to pay Protagoras. Then Protagoras should sue him a second time! Now the student has already won a case, so he should now have to pay Protagoras.

4. The answer is simply that it is logically impossible for there to be such a barber. The given conditions of the problem are simply inconsistent! Look, suppose I told you that a certain man was more than six feet tall and also less than six feet tall—how would you explain that? The answer is that I must be either mistaken or lying.

5. The logician said, "I will be drowned." I leave the proof to the reader that they cannot carry out their decree.

CHAPTER VII
INFINITY AND INDUCTION

I. Infinity

In a certain strange universe, the planets are able to send messages to other planets via space warps. The following facts are known:

1. Each planet has sent a message to at least one other planet.

2. No planet has received messages from more than one planet.

3. There is one planet that has never received any messages at all.

On one of the planets, two astronomers were discussing how many planets there were in the universe. According to one of them, there were less than one hundred trillion. According to the other astronomer, there were more than one hundred trillion.

Problem 1. Which of the two astronomers was right?

Just what is meant by the word *infinite*? Some might reply "endless," although this is not really satisfactory, since the circumference of a circle is certainly endless (it has no beginning or end), yet it would hardly be called *infinite*! As the term *infinite* is used in mathematics and mathematical logic, the word *infinite* is used to describe certain types of *sets*, or *collections*, or *aggregates* of objects of any sort. The whole idea is based on the notion of a 1 to 1 *correspondence*. For example, suppose that in a certain theater every seat is taken and no one is standing (and no one is seated on anybodys lap). Then without having to count the number of seats, or the number of people in the theater, we know that the two numbers are the same, because the set of people is in a 1 to 1 correspondence with the set of seats—each person corresponding to the seat on which he or she is sitting.

More generally, for any two sets A and B, by a 1 to 1 correspondence between A and B is meant a pairing between each element x in A and an element y in B such that each x in A is paired with one and only one element y of B, and for each y in B, there is one and only element x in A that is paired with y.

Consider now the natural numbers—0 and the positive whole numbers 1, 2, ..., n, ... Given a positive whole number n, just what does it mean for a set A to have exactly n elements? It means that it can be put into a 1 to 1 correspondence with the set of positive integers 1 through n. Thus, e.g., to say that there are five fingers on my right hand is to say that I can put the set of fingers of my right hand into a 1 to 1 correspondence with the set {1, 2, 3, 4, 5}—for example, I can pair my thumb with the number 1, my index finger with 2, my middle finger with 3, the next finger with 4, and the pinky with 5. The act of doing this, has a popular name—it is called *counting*. Now, what about the number 0? What does it mean that a set has 0 elements? Obviously it means that it has no elements at all—it is the so-called *empty set*. [I like to characterize the empty set as the set of all people in a theater after everyone has left. That really conveys the feeling of emptiness, doesn't it!]

Now that we know what it means for a set to have n elements (n being a natural number) we define a set A to be *finite* if there is a natural number n such that A has (exactly) n elements; otherwise we say the set is *infinite*. Thus a set A is infinite if and only if there is no natural number n such that A has exactly n elements. An obvious example of an infinite set is the set of all natural numbers.

Are all infinite sets of the same size, or do they come in different sizes? Before we can answer that question we must define exactly what we mean by saying that two sets are of the same size. Well that is easy: We say that set A is the same size as set B if A can be put into a 1 to 1 correspondence with B. Now, what is meant by saying that B is of *larger* size than A? This is trickier! First of all we say that a set S_1 is a *subset* of a set S_2—in symbols $S_1 \subseteq S_2$—if every element of S_1 is also an element of S_2. For example, the set E of even natural numbers is a subset of the set N of all natural numbers. We say that S_1 is a *proper subset* of S_2 if S_1 is a subset of S_2 but not the *whole* of S_2—i.e., there are elements of S_2 that are not in S_1. Thus, e.g., the set E of even numbers is not only a subset of the set N of all natural numbers, but is a *proper* subset of N, since not every natural number is even. Well, one may be tempted to define a set B to be numerically *larger* than a set A if A can be put into a 1 to 1 correspondence with some *proper* subset of B. Well, this definition works fine for *finite* sets, but is no good for infinite sets, because it can well happen that there be two infinite sets A and B such that A can be put into a 1 to 1 correspondence with a proper subset of B and also B can be put into a 1 to 1 correspondence with a proper subset with A. For example, the set E of all even natural numbers can be put into a 1 to 1 correspondence with a proper subset of the set O of all odd natural numbers, but also O can be put into a 1 to 1 correspondence with a proper subset of E as follows: Pair each even number x with the odd number x + 5. Thus 0 corresponds to 5, 2 to 7, 3 to 9, and so forth. Thus E is in 1 to 1 correspondence with the set 5, 7, 9, ..., which is a *proper* subset of O. On the other hand we can put O into a 1 to 1 correspondence with a proper subset of E by pairing each odd number x with the even number x + 5—thus 1 corresponds to 6, 3 with 8, 5 with 10, and so forth. Thus under this correspondence, O is in 1 to 1 correspondence with the set of all even numbers equal to or greater than 6, which is of course a *proper* subset of E. Thus E can be put into a 1 to 1 correspondence with a proper subset of O, and O can be put into a 1 to 1 correspondence with a proper subset of E, yet we certainly would not want to say that E is numerically larger than O and O is also numerically larger than E! Thus the aforementioned proposed definition of "larger than" is not good for infinite sets.

For infinite sets, the proper definition is this: We will say that B is numerically *larger* than A if A can be put into a 1 to 1 correspondence with a proper subset of B, but *cannot* be put into a 1 to 1 correspondence with the whole of B! In other words, A can be put into a correspondence with some subset of B, but given *any* 1 to 1 correspondence from A to some subset B_1 of B, the set B_1 is not the whole of B.

The father of the mathematical field known as *Set Theory* was Georg Cantor whose most famous theorem (known as *Cantor's Theorem*) is that for any set A, there exists a set B that is numerically larger than A! We will soon give a proof of this remarkable result.

A set is called *denumerable* if it can be put into a 1 to 1 correspondence with the set of all positive integers. Cantor first considered whether or not every infinite set is denumerable. As I understand it, Cantor spent twelve years trying to prove that every infinite set is denumerable, and in the thirteenth year, discovered a counter-example (which I like to call a *Cantor-example*). What he did during those first twelve years is to consider various infinite sets which on the surface appeared to be non-denumerable (not denumerable), but then found clever ways to enumerate them—i.e., put them into a 1 to 1 correspondence with the set of all positive integers. I like to illustrate these clever devices as follows:

Imagine that you and I are immortal. I write down a positive integer n on a piece of paper and fold it up and tell you that every day you have one and only one guess as to what the number n is. If and when you guess it, I give you a grand prize. Is there a strategy that will enable you to be sure that you will get the prize sooner or later? Of course there is! Today you ask, "Is it 1"? Tomorrow you ask, "Is it two"? The next day "three," and so forth. Sooner or later you are bound to hit my number.

Now instead, I give you a slightly more difficult test: This time I write down some positive whole number +1, +2, +3, ..., or some negative whole number –1, –2, –3, ..., and again you have one and only one guess each day as to what I have written; and if you guess correctly you get the prize.

Problem 2. What strategy will enable you to be sure of winning the prize sooner or later?

You see from the solution of the above problem that the set of positive and negative whole numbers together, which on the surface seems to be twice as large as the set of positives alone, is really the same size as the positive alone. This illustrates an interesting fact about infinity—namely that it is possible for an infinite set A to be put into a 1 to 1 correspondence with a *proper* subset of itself! No finite set can have that curious property!

Next, I give you a still more difficult task: I write down *two* positive whole numbers, or one such number twice. Again each day you have one and only one guess as to what I have written, and if and when you guess correctly, you win a prize. Is there now a strategy that will guarantee that sooner or later you will win the prize? The situation may well seem hopeless, since there are infinitely many possibilities for the first number that I wrote, and with each of these, there are infinitely many possibilities for the second number! Yet there is a strategy that works.

Problem 3. What strategy would work?

The next test is only a bit more different than the last: I now write down a fraction a/b, where a and b are positive integers. Again you have one and only one guess each day as to what fraction I wrote.

Problem 4. What strategy would enable you to be sure of sooner or later naming the fraction I wrote?

As seen from the solution of the above problem, the set of fractions is indeed a denumerable set—it *can* be put into a 1 to 1 correspondence with the set of all positive integers. This fact, discovered by Cantor, amazed the entire mathematical world at the time!

The next test is more difficult: This time I write down some *finite* set of positive integers. You are not told how many integers are in the set, nor what is the highest number in the set. Nonetheless, there is a strategy that will ensure that you get the prize sooner or later.

Problem 5. What strategy will work?

Cantor's Theorem. We have now seen that the set of all *finite* sets of positive integers is denumerable. What about the set of *all* sets of positive integers—is that set also denumerable? Cantor's great result is that it is *not*! We now turn to this.

Until further notice, the word *number* shall mean positive integer. Let us now imagine a book with denumerably many pages—Page 1, Page 2, Page 3, ..., Page n, ... On each page is described a set of numbers. Is it possible that *every* set of numbers is listed on some page or other? We wish to prove Cantor's result that it is not possible.

Problem 6. Describe a set of numbers that cannot be described on any of the pages. Hint: Call a number n *extraordinary* if n is a member of the set described on Page n. Call n *ordinary* if n is not a member of the set described on Page n.

Is it possible that the set of all extraordinary numbers is listed (described) on some page? Is it possible that the set of ordinary numbers is listed on some page?

Discussion. As is commonly done, I denote by N here the set of positive integers, which are sometimes called the natural numbers. We have now seen that the set of all subsets of N—this set is called the *power set* of N, and is symbolized $\mathscr{P}(N)$—is numerically larger than N. Cantor raised the following problem: Does there exist a set S of intermediate size between N and $\mathscr{P}(N)$? That is, is there a set S which on the one hand is numerically larger than N, but is not as numerically large as $\mathscr{P}(N)$? This is known as the *Continuum Problem*, and has not been solved to this day! Cantor conjectured that there was not such an intermediate set, and this conjecture is known as the *Continuum Hypothesis*, but whether it is true or not remains a mystery. Some, including the author, regard this as the Grand Problem of Mathematics.

A Cantorian Paradox

Now for a few more paradoxes. Let us reconsider the book with infinitely many pages, on each page of which is listed (described) a set of positive integers. We proved that the set of ordinary numbers (numbers n such that n does not belong to the set listed on page n) cannot be listed on any page. But suppose on Page 13 is written: "The set of all ordinary numbers," or "The set of all numbers n such that n does not belong to the set listed on page n." Are we not now in a terrible quandary? The number 13 can be neither ordinary or extraordinary without contradiction. How do we get out of this one?

Problem 7. Yes, how do we?

A Worse One. This should be read only after the solution of the last problem.

To make life more complicated, consider another book with infinitely many pages, but this time each page contains either a genuine description of a set of positive integers or a pseudo-description. We now let S be the set of all numbers n such that the description on Page n is genuine and n does not belong to the set so described. [If the description on Page n is not genuine, then n automatically does not belong to the set S.] This description is obviously genuine. Now suppose that that very genuine description is on Page 13. Is 13 in that set or not? Again, either way we get a contradiction.

Problem 8. What is the solution?

Hypergame and Cantor's Theorem. Again let us consider an infinite sequence $S_1, S_2, ..., S_n, ...$ of sets of positive integers. [We really don't need the book, which is only a cosmetic prop. The important thing are the sets. If desired, one can think of S_n as the set listed on page n]. The set S of all numbers n such that n does not belong to the set S_n is known as *Cantor's diagonal set* (for the sequence $S_1, S_2, ..., S_n, ...$) As we now know, this set S is different from all the sets $S_1, S_2, ..., S_n, ...$ Now, the mathematician William Zwicker came up with a completely different set Z that is different from all the sets S_1, S_2, S_n. He came upon it after having first invented a game called *Hypergame*, which I will now describe.

We consider only games played by two players. Call a game *normal* if it has to terminate in a finite number of moves. Tic Tac Toe is obviously normal. Chess is normal, because of the 50 minute rule. Most games that you are familiar with are normal.

Hypergame is the following: The first move in hypergame is to decide what *normal* game should be played. Suppose you and I are playing hypergame and I am to make the first move. Then I could say, "Let's play chess" and you then make the first move in chess. Or I could say, "Lets play Tik Tak Toe," and you make the first move in Tik Tak Toe. I can choose any normal game I want, but I cannot choose a game that is not normal. Now, is Hypergame normal or not? Well, since I choose a normal game, that normal game must terminate in some finite number n of moves, and so that hypergame terminates in n + 1 moves, hence Hypergame must be normal. Now that Hypergame is normal, on my first move I can say, "Let's play Hypergame," Then you say, "Let's play Hypergame." Then I can say, "Let's play Hypergame" .... And so this could go on forever, which means that Hypergame is not normal. This is the paradox.

Problem 9. How can this paradox be resolved?

Zwicker's Set. This paradox cleverly suggested to Zwicker the following way of proving Cantor's theorem— he constructed a completely different set S which differs from all the sets S_1, S_2, ..., S_n, ... Here is the construction: For any number x (positive integer, that is), by an x-path we mean a sequence obtained as follows: Go to the set S_x. If that set is empty (contains *no numbers*) that's the end of the path. Otherwise pick some number y from S_x and go to the set S_y. If that set is empty, that's the end of the path. Otherwise pick some number z from S_y and go to the set S_z, and continue in this manner until you either come to the empty set, or the process goes on for ever, in which case the x-path is infinite. Now call x *normal* if all x-paths are finite. Now let Z be the set of all normal numbers (Zwicker's set). This set Z is different from all the sets S_1, S_2, ..., S_n, ...

Problem 10. Prove that the set Z of all normal numbers cannot be any one of the sets S_1, S_2, ..., S_n, ... [Hint: Suppose that set was, say, S_{13}. Is 13 normal or not?]

Once at a lecture I gave in which I presented Zwicker's proof, someone asked me what happens if the empty set is not any of the sets S_1, S_2, ..., S_n, ... Is the set of all normal numbers still different from each of the sets S_1, S_2, ..., S_n, ...?

Problem 11. What is the answer?

II. Mathematical Induction

Let me begin with a story. A certain man was in quest of immortality. He read many occult books on the subject, but none of them gave him any practical advice on how to become immortal. Then he heard of a certain great sage of the East who knew the true secret of immortality. It took him twelve years to find the sage, and when he did, he asked, "Is it really possible to become immortal?" The sage replied, "It is really quite easy, if you do just two things." "And what are they?" the man asked quite eagerly. "First of all," replied the sage, "from now on, you must always tell the truth. You must never make a false statement. That's a small price to pay for immortality, isn't it?" "Of course!" was the reply. "Secondly," continued the sage just say, "I will repeat this sentence tomorrow." If you do these two things, I guarantee you will live forever." After thinking for a few minutes, the man said, "Of course if today I truthfully say that I will repeat this sentence tomorrow, then I will indeed say it again tomorrow, hence again the next day, and the next and the next and so on, but your solution is not very practical. How can I be sure of truthfully saying that I will repeat this sentence tomorrow if I don't know for sure that I will be alive tomorrow? Your advice, though interesting, is simply not practical."

"Oh," said the sage, "You wanted a *practical* solution! No, I deal only in theory."

Of course the sage was theoretically right! If I truthfully say "I will repeat this sentence tomorrow" and remain truthful in the future, then of course I will live forever. In the argument the man gave, the principle of mathematical induction, that I will shortly explain was hidden in the phrase "and so on."

Now for a little problem, which is indeed related to the principle of mathematical induction: We return to the Island of Knights and Knaves in which knights make only true statements and knaves make only false ones. One of the inhabitants once said, "This is not the first time I have said what I am now saying."

Problem 12. Was he a knight or a knave?

Consider now the following situation: On a certain planet it is raining today. Also it never happens that it rains one day and not on the next—on any day that it rains, it also rains the next day as well. Isn't it then obvious that on this planet it will rain forevermore? This is essentially the principle of mathematical induction. In general, the principle is this: If a certain property holds for the number zero, and whenever it holds for a number n, it also holds for n + 1, then it holds for all natural numbers. This is the principle of mathematical induction.

Here is another illustration of the principle: Imagine that we are all immortal and as in the good old days, the housewife leaves a note in the morning in a small bottle telling the milk man what to do. Now suppose she leaves the following note:

NEVER LEAVE MILK ONE DAY UNLESS YOU LEAVE MILK THE NEXT DAY AS WELL.

Well, suppose the milk man never leaves any milk at all. If after a few days the lady complains, the milk man can justly say, "I never disobeyed your order! Did I ever leave milk one day and not the next? Surely not! I never left milk on any day at all!"

The milk man was absolutely right! He never disobeyed the lady's order. Thus the lady's note was simply inadequate to guarantee permanent delivery—indeed, it did not guarantee delivery on any day at all.

The following note would guarantee permanent delivery:

1. LEAVE MILK TODAY.

2. ON ANY DAY THAT YOU LEAVE MILK, LEAVE MILK ON THE NEXT DAY AS WELL.

The computer scientist Alan Tritter once suggested to me the following cute alternative which illustrates an essential idea behind the fields known as *recursion theory* (which we will go into later), as well as the computing devices invented by the British logician Alan Turing that are known as *Turing Machines.*

Alan Tritter's note has only one sentence:

LEAVE MILK TODAY AND READ THIS NOTE AGAIN TOMORROW

That surely guarantees permanent delivery!

Solutions to the Problems of Chapter VII

1. There must indeed be more than a hundred billion planets—in fact in this universe there must be infinitely many planets! Here is why:

There is one planet that has never received any messages at all—call it Planet 0. Planet 0 has sent a message to at least one other planet—call it Planet 1. This Planet 1 has sent a message to at least one *other* Planet. This other planet cannot be Planet 0, which has never received any messages, nor can it be Planet 1, since Planet 1 sent a message to some *other* planet, hence Planet 1 sent a message to some planet other than Planet 0, or Planet 1—call it Planet 2. Now, Planet 2 has sent a message to some other planet. This other planet cannot be Planet 0, which has never received any messages, nor Planet 1, which has received a message from Planet 0, and no planet has received more than one message from another planet, hence the planet that received a message from Planet 2 is different from Planet 0 and Planet 1, and thus is a third planet—call it Planet 3. This Planet 3 has sent a message to some other planet, which cannot be Planet 0 (for same reason as before), nor Planet 1 or Planet 2, which has received a message from Planet 0 or Planet 1, respectively, hence must be some new planet—call it planet 4. We can then go on infinitely, thus forming an infinite sequence Planet 0, Planet 1, ..., Planet n, Planet n + 1, ...

Thus there must be infinitely many planets in this strange universe.

Hidden in the above proof is the principle of *mathematical induction*, which we will deal with in Part II of this chapter.

2. Obviously your strategy is to alternate between the positives and negatives thus: you guess numbers in the order 1, –1, 2, –2, 3, –3, etc.

3. For each number n, there are exactly n pairs whose highest number is n—namely $(1, n)$, $(2, n)$, ... $(n-1, n)$, (n, n). So you start with those pairs whose highest number is 1, then those whose highest number is 2, then those whose highest number is 3, and so forth. Thus you guess the pairs in the order $(1, 1)$, $(1, 2)$, $(2, 2)$, $(1, 3)$, $(2, 3)$, $(3, 3)$, $(1, 4)$, $(2, 4)$, $(3, 4)$, $(4, 4)$, and so forth.

4. For each number n there are only finite many fractions whose highest numerator or denominator is n— exactly $2n-1$, in fact—namely $1/n$, $2/n$, ..., $n-1/n$, n/n, $n/1$, $n/2$, ..., $n/n-1$. You thus enumerate the fractions in the order $1/1$, $1/2$, $2/2$, $2/1$, $1/3$, $2/3$, $3/3$, $3/2$, $3/1$, ...

5. For each number n, there are only finitely many sets whose highest member is n (there are exactly 2^{n-1}, in fact), and so we first go through all sets whose highest number is 1 (there is only one such—the set $\{1\}$ whose only member is 1), then the two sets whose highest number is 2 (the set $\{2\}$, $\{1, 2\}$), then the four sets whose highest number is 3 (the sets $\{3\}$, $\{1, 3\}$, $\{2, 3\}$, $\{1, 2, 3\}$, then the eight sets whose highest member is four, and so forth.

6. There is no reason why the set of extraordinary numbers cannot be listed on one of the pages, but the set of all ordinary numbers cannot be, for if it were, this would be a paradox. Suppose, for example, that the set of all ordinary numbers was listed on page 13. Is 13 then ordinary or not? Either way we get a contradiction: If 13 is ordinary, then it belongs to the set of all ordinary numbers, but this set is listed on page 13, hence 13 belongs to the set listed on page 13, which makes 13 extraordinary by definition. Thus it is a contradiction to say that 13 is ordinary. On the other hand, suppose that 13 is extraordinary. Then by

definition it belongs to the set listed on Page 13, but that is not possible, since only ordinary numbers are in that set. Thus if the set of all ordinary numbers were listed on Page 13, the number 13 couldn't be either ordinary or extraordinary without contradiction. Thus it is not possible that the set of all ordinary numbers could be listed on Page 13.

Of course the number 13 was chosen arbitrarily. For any number n, the set of ordinary numbers could not be listed on Page n, for if it were, the number n itself couldn't be either ordinary or extraordinary without contradiction.

7. The answer is that if "the set of ordinary numbers" is written on any page—say Page 13—the phrase is not a genuine description, since it does specify whether 13 belongs to that set or not. It is what logicians call a *pseudo-description.*

The interesting thing is that if that phrase is written outside the book, it is genuine (assuming all descriptions in the book are genuine), but if it is written inside the book, it fails to be genuine!

8. The answer to this is that the very notion of a genuine description is not well-defined. The situation is similar to the Richard paradox: "The smallest number not describable in less than eleven words."

The above purported description has only ten words!

9. Hypergame is not really well defined. Yes, given a set S of well defined games, one can speak of hypergame *for the set S*, and that is indeed well defined, but that hypergame itself cannot be a member of the set S.

10. Alright, suppose the set of all normal numbers was described on some page—say Page 13. The following paradox would then arise: On the one hand 13 must be normal, because consider any path starting with 13. We go to the set S_{13} (the set described on Page 13) and if this set

is not empty we pick a number x from S_{13}. Thus the path begins 13, x, ... Now x must be normal (since this part of the sequence beginning with x must terminate. But since 13 is normal, it must be a member of S_{13} (because S_{13} is the set of *all* normal numbers), and so I pick 13 out of S_{13}, go back to S_{13}, again pick out 13, and so forth, thus generating the infinite path 13, 13, 13, ..., 13, ..., which makes 13 not normal. Thus it is not possible for S_{13} (or S_n for any other n) to be the set of all normal numbers.

11. The answer is *yes*, because if none of the sets S_1, S_2, ..., S_n, ... are empty, then no number is normal, hence the set of all normal numbers is then the empty set, which is different from all the sets S_1, S_s, ..., S_n, ...

12. If he were a knight, we would get a contradiction because what he said was then true, which means that he really had made that same statement previously. When he made it then he was still a knight, hence he must have said it a time before that. Hence a time before that, and a time before that, and .... Thus unless he has lived infinitely far back in the past, he cannot be a knight. He must be a knave.

Another way to look at it is this: Since he said it once, there must have been a first time he said it, and when he said it then, it was obviously false.

CHAPTER VIII

INTRODUCING SELF-REFERENCE

Self-Reference plays an important part in the theory of mathematics, logic and computer science, as we will later see. In this chapter we consider various methods of achieving self-reference.

Use versus Mention. Consider the following two sentences

(1) Snow is white.

(2) Snow has four letters.

Are both of these sentences true? Sentence (1) is certainly true, but sentence (2) is not; snow doesn't have any letters. It is the *word* "snow" that has letters. And so the correct rendition of (2) is

(2)' "Snow" has four letters.

To *use* a word, is to refer to the thing that the word denotes, whereas to *mention* a word is to refer to the word itself. Thus in Sentence (1), the word "snow" is used, not mentioned. In Sentence (2), the word "snow" is also used, but wrongly so. In Sentence (2)', the word "snow" is not used, but mentioned—and correctly so. Many jokes are based upon a confusion between use and mention. In fact I once witnessed the following: A teacher said to a girl, "Please write your name on the blackboard." The girl went to the blackboard and wrote:

$$\boxed{\text{your name}}$$

She obviously knew that the teacher was *using* "your name," but she mischievously acted as if the teacher was *mentioning* it.

Sometimes a word can be both used and mentioned in the same sentence, such as

"William Shakespeare" is the name of William Shakespeare.

Or as a more humorous illustration:

This sentence is longer than "this sentence."

Or another:

It takes longer to read the Bible than to read "the Bible."

It sure does!

A humorous illustration of a confusion that can arise between an expression and its name was given by Lewis Carroll in the following exchange between the White Knight and Alice in *Alice's Adventures through the Looking Glass*:

White Knight	The name of the song is called *"Haddock's Eyes."*
Alice	Oh, that's the name of the song, is it?
White Knight	No, you don't understand that's what the name is *called*. The name really is "The Aged Aged Man."
Alice	Then I ought to have said "That's what the song is called?"
White Knight	No, you oughtn't: that's quite another thing! The *song* is called *Ways and Means* but that's only what it's *called*, you know!
Alice	Well, what *is* the song then?
White Knight	I was coming to that. The song really is *"A-sitting on A Gate"*.

Here, 1 believe, Lewis Carroll made an error: He made a distinction between the name of the song and what the song is called. The two are really the same thing. So Alice was actually right when she said, "Then I ought to have said 'That's what the song is called!'"

There are many ways in which an expression can be named. A common way is by putting quotation marks around it. Thus in the expression "snow", putting quotes around the sequence of 4 letters, s, n, o, w has named the word formed by those letters.

Consider the following sentence:

(1) " "Snow" " has two pairs of quotation marks.

Is that sentence true or false? Many will wrongly say that it is true. It is not true! What is *written* prior to "has two pairs of quotation marks" does indeed have two pairs of quotation marks, but what it denotes has only one pair of quotation marks. And so the correct version is

(1)' " "Snow" " has one pair of quotation marks.

Self-Reference without an Indexical. Suppose we wish to write a sentence that asserts that John is reading that very sentence. One way is this:

JOHN IS READING THIS SENTENCE.

Now, the above sentence uses the indexical word "this." As many of you know, an indexical word is one whose meaning (denotation) depends on where or when it occurs. For example, "I" denotes different people when uttered by different people. The indexical word "now" denotes different times when uttered at different times.

Now, this and several subsequent chapters are leading us to Gödel's famous theorem that for mathematical systems of sufficient strength, there must always be true sentences that are not provable in the system [Gödel, 1931]. The systems studied by Gödel had no indexicals, and so we now wish to do for ordinary English what Gödel did for mathematical systems—that is we now wish to construct a sentence that asserts that John is reading it, but which does not use any indexical. We now turn to a few different ways of doing this.

Diagonalization. Consider the following expression:

(1) John is reading x.

That expression is not a sentence, since we don't know what the letter "x" stands for. It is what is called a *sentence form*, and becomes a sentence only after one replaces the letter "x" by the name of some expression. Then it becomes either true or false. The letter "x" is here used as what is known as a *variable*, which stands for some unknown expression.

Consider now any expression in which the variable "x" occurs. By its *diagonalization* is meant the result of replacing the variable x by the quotation of the entire expression. For example, the diagonalization of (1) is

(2) John is reading "John is reading x."

Now, (2) is not self-referential; it does not assert that John is reading (2); it asserts that John is reading (1). But now consider the following expression:

(3) John is reading the diagonalization of x.

The expression (3) is not a sentence; it is a sentence form. Its diagonalization is

(4) John is reading the diagonalization of "John is reading the diagonalization of x."

Now, (4) asserts that John is reading the diagonalization of (3), but the diagonalization of (3) is (4) itself, hence (4) asserts that John is reading the very sentence (4)!

Perhaps this will be more clear if we use the following abbreviations: Let the letter "J" abbreviate "John is reading," and let the letter "D" abbreviate "the diagonalization of." Then the symbolic form of (3) is

(5) JDx (read "John is reading the diagonalization of x")

Its diagonalization is

(6) JD "JDx"

(6) asserts that John is reading the diagonalization of (5), which is (6) itself!

Normalization. In my paper "Languages in which Self-Reference is Possible" I introduced another method of achieving self-reference which in application has turned out to have certain advantages. Unlike diagonalization, it avoids the operation of substituting the name of an expression for a variable "x." In fact it uses no variables.

By the *norm* of an expression I mean the expression followed by its own quotation. For example, consider the following expression:

(7) John is reading.

The norm of (7) is

(8) John is reading "John is reading." The expression (8) is a sentence and it asserts, not that John is reading (8), but that John is reading (7). Thus (8) is not self-referential.

But now consider the following:

(9) John is reading the norm of

Its norm is:

(10) John is reading the norm of "John is reading the norm of." The sentence (10) asserts that John is reading the norm of (9), which is the very sentence (10).

Again it will be more perspicuous if we abbreviate "John is reading" by "J," and "the norm of" by "N." Then (9) and (10) symbolically are

(11) JN

(12) JN "JN"

Sentence (12) asserts that John is reading the norm of (11) which is (12) itself.

Self-Reference using Gödel numbering. Gödel achieved self-reference by assigning to each expression of the language he was considering a number called the *Gödel number* of the expression, which somehow served the role of a name of the expression. There are simpler Gödel numberings than the one Gödel used, and in this chapter I will be using one of mine which employs binary numbers. We continue to use the symbols J and N, but instead of quotation marks we now use the two symbols 1 and 0. To the four symbols, J, N, 1, 0, we assign numbers as follows:

J—10

N—100

1—1000

0—10000

To any compound expression built from their four symbols (whether meaningful or not) we will take the Gödel number of the expression to be the result of replacing each symbol by its Gödel number. For example the Gödel number of JJ is 1010, the Gödel number of 1N is 1000100; the Gödel number of NJ0 is 1001010000. We now redefine the *norm* of an expression to be the expression followed, not by its quotation, but by its Gödel number. For example, the norm of NOJ is NOJ1001000010 (since 1001000010 is the Gödel number of NOJ).

By a *sentence* we should now mean any expression of the form JNg, where g is any Gödel number—that is, a sentence consists of the letter J followed by the letter N followed by a binary number which is the Gödel number of some expression. The sentence JNg is interpreted to mean that John is reading the norm of that expression whose Gödel number is g.

Problem. Using the four symbols J, N, 0, 1, construct a sentence that asserts that John is reading it.

Comments. The idea of Gödel numbering has been used for a well known joke about the numbering of jokes. The joke has two endings, one of which is well known, and the other is much less well known, but I believe you will agree that it is much funnier! I will tell you both endings.

The joke is about a man who was president of a joke makers club who invited a friend to a banquet of the joke-makers. At the banquet the friend was puzzled by the proceedings. At various times, someone would get up, call out a number and everyone would laugh. The friend asked the host what was going on. The host then explained, "We joke-makers don't want to take the time and trouble to tell the whole joke, and so we number our jokes. Someone gets up and calls out the number of a joke and this recalls the joke to mind, and so everyone laughs."

Now, the well known ending is this: The friend asks, "Can I try it?" The host agrees. The friend then calls out a number, but no one laughs. The friend then asked the host what was wrong. The host replies, "Some people can tell a joke, and some people can't!"

That is the well known ending. The other ending, which I like much better is this: At one point of the banquet, someone got up and called a very high number. Everyone laughed and one man kept laughing for quite a time after the others had finished. The guest then asked the host, "Why is he still laughing?" The host replied, "Oh, he hasn't heard that one before."

Solution to the Problem of Chapter VIII

The sentence is JN10100. It asserts that John is reading the norm of that expression whose Gödel number is 10100, but the expression whose Gödel number is 10100 is JN, and so the sentence asserts that John is reading the norm of JN, which is the very sentence JN10100.

CHAPTER IX
FIXED POINT PUZZLES

Fixed points (you'll find out what they are) are closely related to Self-Reference, which in turn is closely related to Gödel's great theorem, which we are approaching very gradually!

§1. Arnie's System

Part 1

I have a friend Arnold who invents all sorts of curious things. On one of my visits to his workshop, he showed me a strange looking gadget he had recently invented. "I have been having a lot of fun with this device," said Arnie. "It does all sorts of interesting things."

"Does it have any practical applications?" I innocently asked.

"I have no idea!" he angrily replied, "and I couldn't care less! Why are people always demanding practical applications? Utility is fine, but aren't there other values in life? What about aesthetic and spiritual values; don't they count? To me, the main value of Mathematics and Logic is not utility, which of course is important, but the aesthetics of the presentations and of the results themselves! I think of Mathematics as a fine art!"

I thought it wise not to press this further, and so I asked him to explain to me the function of his gadget.

"This device operates on expressions consisting of strings of capital letters," he said. "Any string of capital letters is called an *expression*. An expression doesn't have to be meaningful. For example BLPQ is an expression and so is AHM—also any capital letter itself is an expression. Now, one feeds an expression into the gadget, and if the expression is acceptable, the machine then outputs an expression—usually different from the input. I shall use small letters x, y, z as standing for arbitrary expressions (much as in algebra one uses x, y, z as standing for unknown numbers). And for any pair of expressions x and y, by xy I mean the expression x followed by the expression y—for example, if x is the expression BLD and y is the expression HIJZ, then xy is the expression BLDHIJZ. Is that clear?"

"Yes," I said, "Go on!"

"I shall say that an expression x *yields* an expression y—in short 'x yields y'—to mean that if x is fed in as input, y comes out as output."

Arnie then told me two rules concerning the operation of the gadget. He first mentioned his "Rule Q:"

Rule Q. For any expression x, the expression Qx yields x.

"For example, QBLZ yields BLZ," said Arnie. "Let me show you."

He then fed in QBLZ, and sure enough, out came BLZ.

"Before I state the second rule," said Arnie, "I must tell you that for any expression x, the expression xQx bears a particularly significant relation to x, and is known as the *companion* of x. My second rule is this:

"*Rule C*. If x yields y, then Cx yields the companion yQy of y.

"For example, since QAB yields AB, then CQAB should yield the companion of AB—the expression ABQAB—here, let me try it."

Arnie then fed in CQAB, and sure enough, out came ABQAB.

"My gadget never fails!" said Arnie, with a proud smile. "Indeed, for any expression x, CQx will yield the companion of x. Now, let me try you on some problems." He handed me a sheet of paper with the following four problems written on it.

Problem 1. Find an x that yields itself (you can feed in x, and x comes out).

Problem 2. Find an x that yields its own companion.

Problem 3. Find an x that yields the companion of its companion.

Problem 4. Prove that for any expression y, there is some x that yields yx. For example, what x yields Ax?

I started working on the four problems and when I had finished them, I felt I was beginning to get the hang of how his system worked.

"There are other rules for my gadget," said Arnie, after nodding when I presented him with my solutions. "For any expression x, by the *repeat* of x I mean xx—that is x followed by itself. For example, the repeat of HJK is HJKHJK. Here is another rule governing my device."

"Rule R. If x yields y, then Rx yields yy—the repeat of y."

"With this additional rule there are many more things that the device can do. Here are some problems to help you understand the new Rule R." Arnie handed me a sheet containing the following problems, which took me some time to solve.

Problem 5. Find an x that yields its own repeat.

Problem 6. Find an x that yields the repeat of its companion.

Problem 7. Find an x that yields the companion of its repeat.

Problem 8. Find an x that yields the repeat of its repeat—i.e., one that yields xxxx.

Problem 9. We have already found an x that yields itself, and this x is composed of the letters Q and C. There is another expression that does not use the letter C and which yields itself. What expression does this?

Problem 10. We know that RQx yields the repeat of x. Is there some x such that Rx yields the repeat of x?

Problem 11. We have seen that CQAC is an x that yields Ax (CQAC yields ACQAC). There is another expression x using just the letters Q, R, and A that yields Ax. What x does this? More generally, given any expression y, there is some x that uses just the letters Q, R and the letters of y that yields yx. What x does this?

"Now for another rule," said Arnie, when I had finished solving the problems. "By the *reverse* of an expression is meant the expression with its letters in reverse order—for example the reverse of LXQM is MQXL.

"*Rule V.* If x yields y, then Vx yields the reverse of y.

"With this new rule, several more constructions are possible." Then Arnie handed me another sheet of problems, including an open problem that he himself couldn't solve. In addition there were two new definitions!

Problem 12. Find an x that yields its own reverse.

Problem 13. Find an x that yields the reverse of its repeat.

Problem 14. Find an x that yields the reverse of its companion.

Problem 15. Find an x that yields the companion of its reverse.

Problem 16. Find an x that yields the repeat of the reverse of its companion.

Problem 17. It can be proved that there is only one expression composed of just the letters Q and C that yields itself. However there is more than one expression using the letters Q, C and V that yields itself. How many are there? Also, how many self-yielding expressions are there that use just the letters Q, R and V?

For any expression x, I use the notation $\overleftarrow{x}$ to mean the *reverse* of x.

Problem 18. Prove that for any expression y there is some x that yields $\overleftarrow{x}$y. For example, which expression x yields $\overleftarrow{x}$JKL?

An Open Problem. Given an expression y, is there necessarily an x that yields xy? For example, is there an x that yields xA? I wish I knew the answer! If any of you can prove or disprove it, please let me know.

Symmetric Expressions. We call an expression *symmetric* if it reads the same forwards and backwards—in other words if it is equal to its own reverse. Now, except for the symmetric expressions CQC, which yields itself, the other self-yielding expressions given in the solution of Problem 17 are not symmetric.

Problem 19. Find some symmetric expressions other than CQC that yield themselves. How many of them are there?

This time it took me quite a while to solve Arnie's problems, but when I had finished he immediately started telling me about even more rules.

"Here is another rule I have been having fun with," said Arnie. "I will still use the notation $\overleftarrow{x}$ to mean the reverse of x.

"Rule P. If x yields y, then Px yields the symmetric expression y$\overleftarrow{y}$.

"For example, if x yields ABC, then Px yields ABCCBA.

"I use the symbol 'P,' said Arnie, to suggest the word palindrome. A symmetric expression could also be called a palindrome, since it reads the same backwards and forwards. I call $x\overleftarrow{x}$ the *palindrome* of x." Arnie then passed me three palindrome problems, which I finished off quickly.

Problem 20. Find an x that yields its own palindrome $x\overleftarrow{x}$.

Problem 21. Find an x that yields $\overleftarrow{x}x$.

Problem 22. Find an x that yields the palindrome of its companion.

"Here is still another rule I have been having fun with," said Arnie, picking up as soon as I turned to him. "For any expression x of length two or more, by $\dot{x}$ I mean the result of moving the first letter of x to the end of x—for example, if x is the expression KGS, then $\dot{x}$ is GSK. If x consists of only one letter, then I take $\dot{x}$ to be x itself.

"Here is my next rule:

"Rule M. If x yields y then Mx yields $\dot{y}$.

"I use the letter M to suggest the word *move*. The expression $\dot{y}$ is the result of moving the beginning letter of y to the end." Arnie then passed me a sheet that contained two more problems.

Problem 23. Find an x that yields $\dot{x}$.

We left as an open problem whether from just Rules Q, C, R and V it can be determined if for every y there is some x that yields xy. Now with the addition of Rule M, it is relatively easy to find an x that yields xy, for any y.

Problem 24. How? For example, what x yields xJKL?

Arnie next explained to me that most of the previous results were but special cases of one very lovely general fact involving the important concept of *Fixed Points*:

"I shall use the symbol Θ to denote any expression having the property that for every expression x, the expression ΘQx yields some expression y, and for any z other than Qx that yields x, Θz yields the same y as is yielded by ΘQx. Such an expression Θ I will call a *special* expression. Thus for any special expression Θ, if y_1 and y_2 yield the same expression, then $Θy_1$ and $Θy_2$ will yield the same expression. Any expression composed of the letters C, R, and V is special. For any special expression Θ and any expression x, by Θ(x) I shall mean the expression yielded by ΘQx (or by Θy for any y that yields x). For example R(x) is the repeat of x (since RQx yields xx); V(x) is the reverse of x; C(x) is the companion xQx of x; RC(x) is the repeat of the companion of x; CV(x) is the companion of the reverse of x. Is this clear so far?"

"Yes," I said.

"Good. Now I call x a *fixed point* of Θ if x yields Θ(x). Many of the previous problems were tantamount to finding fixed points. For example, finding an x that yields its own repeat is finding a fixed point of R; an x that yields its own reverse is simply a fixed point of V; an x that yields the repeat of the companion of x is a fixed point of RC, and so forth. However, we can neatly generalize all this."

Arnie then stated the following principle, which I will call *Arnie's Fixed Point Principle.*

Principle AF. Every special expression Θ has a fixed point.

Arnie continued, "Even if I add more rules so that more expressions become special, my fixed point principle will still hold. Any system using Rules Q and C, plus any other rules, obeys my principle. Also, any system using Q and R and any other rules, obeys the principle. That is, for any special Θ, there is a fixed point of Θ that uses just the letters Q and C, or Q and R, plus any letters in Θ.

"Here's an important problem," Arnie said, handing me a sheet of paper with only one problem on it, after which he adjourned to his kitchen to find us some liquid refreshments while I solved it:

Problem 25. Prove Principle AF. Given Θ, what expression is a fixed point of Θ? There are two solutions—one uses just the letters Q and C and the letters in Θ; the other uses just the letters Q and R plus the letters in Θ.

Doubling Up. "My system has another interesting property," said Arnie, after handing me a lemonade. "Now try the following problem."

Arnie then handed me the following problem, which unfortunately I was unable to solve, though it indeed has a solution!

Problem 26. Find two expressions x and y such that x yields the repeat of y and y yields the reverse of x.

"Now how did you ever find those x and y?" I asked, after Arnie showed me the solution.

"It's quite simple," laughed Arnie, "after you see my double fixed point principle. Indeed once one realizes this principle, one can do all sorts of things—for example find an x that yields the palindrome of the companion of y, while y yields the reverse of the repeat of x—and many other things like it."

"I am all ears," I said. "What is this amazing principle?"

Arnie smiled as he showed me the following double fixed point principle (which I call "Arnie's Double Fixed-Point Principle"):

Principle ADF. For any pair of special expressions Θ_1 and Θ_2, there are expressions x and y such that x yields $\Theta_1(y)$ and y yields $\Theta_2(x)$.

"How in the world do you prove this?" I asked.

"Let me lead up to this gradually," he replied. "First try the following problem." He handed me another paper and pointed at the first problem on it.

Problem 27. Using just Rules Q and C, find two *distinct* expressions such that each one yields the other.

At first I didn't know how to approach this, but then, by looking back at earlier problems I had solved at the very beginning, I saw the solution. Arnie showed me how pleased he was at how quickly I had solved the problem, and continued.

"Now for a key problem," said Arnie, pointing at the second problem on the sheet.

Problem 28. Show that for any expressions z_1 and z_2 there are expressions y_1 and y_2 such that y_1 yields $z_1 y_2$ and y_2 yields $z_2 y_1$. For example, what expressions y_1 and y_2 are such that y_1 yields $A y_2$ and y_2 yields $B y_1$?

I spent a few minutes solving this problem, and when I was done Arnie said, "Now you should easily be able to prove my double fixed point principle." He pointed at the last two problems on the sheet. I found them to be an enjoyable challenge.

Problem 29. Prove the double fixed point principle.

Problem 30. Find two *distinct* expressions x and y such that each yields the repeat of the other.

§2. Another System

But Arnie wasn't done with me yet! "Remember," Arnie said later, "in your book THE LADY OR THE TIGER you gave a very curious system in which the only expression that you could find that yielded itself was of length 10. Your system had four rules—Rules R and V are the same as mine, but instead of my Rules Q and C, you had the following two Rules:

"*Rule QQ.* QxQ yields x.

"*Rule L.* If x yields y, then Lx yields Qy.

"You stated without proof that your system obeys the fixed point principle. How would you prove this? For example, how do you find a fixed point of R—an expression that yields its own repeat? Related to this, is it true that as in my system, for any expression y there is some x that yields yx? You also stated without proof that your system obeys the double fixed point principle. How do you prove that? For example, how do you find expressions x and y such that x yields the repeat of y and y the reverse of x?"

And then Arnie handed me six problems about my very own system! He asked for my advice on how to solve them.

Problem 31. Find an expression x that yields its own repeat in the "Lady or the Tiger" system.

Problem 32. Is it true that in this system, for any expression y there is an expression x that yields yx?

Problem 33. How do you prove the fixed point principle for this system?

Problem 34. Find expressions x and y such that x yields the repeat of y and y yields the reverse of x.

Problem 35. I suspect that it is true that in this system, for any z_1 and z_2 there is some y_1 and y_2 such that y_1 yields $z_1 y_2$ and y_2 yields $z_2 y_1$? But how is this proved?

I realized that Problem 35 isn't easy. I gave Arnie the following hints regarding solving it: First find an expression—call it Δ—such that if x yields y, then Δx yields yQyQQ. Then show that for any expression y there is some expression z that yields yQzQ.

Arnie's Problem 35 was the principal challenge, but not that difficult after the other problems were solved regarding my system. We spent most of the rest of the afternoon working through these problems together. And then I came up with one further problem for him.

Problem 36. How do you prove that the double fixed point principle holds for this system?

Solutions to the Problems of Chapter IX

I would like to sometimes use the notation x → y to abbreviate "x yields y."

1. Since Qx → x, then CQx → xQx. We take C for x, and we have CQC → CQC, and so CQC yields itself.

2. Since CQx yields the companion of x, then CCQx yields the companion of the companion of x. We now take CC for x, and so CCQCC yields the companion of the companion of CC, but since the companion of CC is CCQCC, then CCQCC yields the compaion of CCQCC. Thus CCQCC yields its own companion.

3. CCCQCCC yields the companion of its companion, as the reader can verify.

4. CQAC yields the companion of AC, which is ACQAC, and thus CQAC yields A followed by CQAC. Indeed, for any expression y, an x that yields yx is CQyC.

This fact will later prove very important, as the reader will find out.

5. For any x, the expression CQx yields the companion of x. In particular (taking RC for x), CQRC yields the companion of RC, which is RCQRC. Thus RCQRC yields the repeat of RCQRC. Thus RCQRC yields its own repeat.

6. RCCQRCC

7. CRCQCRC

8. RRCQRRC

9. RQRQ yields itself. [Do I hear any RQ-ment?]

10. Any x that yields itself will do the trick—for example, CQC, or RQRQ.

11. RQARQ yields ARQARQ. Thus for x = RQARQ, the expression x yields Ax. More generally, for any y, we take x to be RQyRQ, and x yields yx.

12. VCQVC

13. VRCQVRC

14. VCCQVCC

15. CVCQCVC

16. RVCCQRVCC

17. Let us first note that if x yields y, so does VVx, since it yields the reverse of the reverse of y, which is y itself. Also VVVVy yields y, and so does zx yield y, where z is any expression consisting of an even number of Vs. For any such expression z, the expression zCQzC will yield itself, because CQzC yields the companion of zC, which is zCQzC. Hence zCQzC also yields zCQzC. Thus there are infinitely many expressions that yields themselves.

There are also infinitely many self-yielding expressions using the letter R instead of C. For instance, an expression of the form zRQzRQ, where z consists of an even number of R's, will do.

18. An x that yields $\overleftarrow{x}$JKL is VCQLKJVC, because CQLKJVC yields the companion of LKJVC, which is LKJVCQLKJVC. Therefore VCQLKJVC yields the reverse of LKJVCQLKJVC, which is CVJKLQVCJKL. Thus VCQLKJVC yields $\overline{VCQLKJVC}$JKL.

More generally, for any expression y, an x that yields $\overleftarrow{x}$y is VCQ$\overleftarrow{y}$VC.

19. VVCVVQVVCVV is a symmetric self-yielding expression. So is zCzQzCz, where z consists of an even number of Vs.

20. PCQPC

21. VPCQVPC

22. PCCQPCC

23. MCQMC

24. If y consists of just one letter, then an x that yields xy is MCQyMC, because CQyMC yields yMCQyMC, hence MCQyMC yields MCQyMCy. If y is of length 2, then an x that yields xy is MMCQyMMC. In general, for any y, if we

IX: Fixed Point Puzzles 97

let z be a string of M's of the same length as that of y, an x that yields xy is zQyzC.

25. If x is any expression that yields Θx, then Θx will yield $\Theta(\Theta$x), and thus Θx will be a fixed point of Θ. Well, we know that an x that yields Θx is CQΘC (solution of problem 4), and so ΘCQΘC is a fixed point of Θ. Another x that yields Θx is RQΘRQ, and so ΘRQΘRQ is also a fixed point of Θ.

26. Take x = RQVCQRQVC,
           y = VCQRQVC.

How x and y were found will soon be clear.

27. What is required is an expression x that yields Qx. In turn, Qx yields x. Well, we know from the solution to Problem 4 that an x that yields Qx is CQQC. And so the sentences CQQC and QCQQC are distinct and each one yields the other.

28. There are several ways to find expressions y_1 and y_2 such that y_1 yields Ay_2 and y_2 yields By_1. One way is to first find y_1 such that y_1 yields AQBy_1 and then take y_2 to be QBy_1. Thus y_1 yields Ay_2 (which is AQBy_1) and of course y_2 (which is QBy_1) yields By_1. There are two ways of finding such an expression y_1—either using Q and C (as in the solution of Problem 4) or using Q and R (as in the solution of Problem 11). The first way yields a shorter expression, and is the one we will adopt. Thus we take y_1 and y_2 as follows.

y_1 = CQAQBC,
y_2 = QBCQAQBC.

Alternatively, we could first find some y_2 that yields BQAy_2 and take y_1 to be QAy_2. This gives the solution:

y_1 = QACQBQAC,
y_2 = CQBQAC.

We shall adopt the first pair of alternatives, and so for arbitrary expressions z_1 and z_2, we should take y_1 and y_2 as follows:

$y_1 = CQz_1Qz_2C,$
$y_2 = Qz_2CQz_1Qz_2C.$

Then y_1 yields z_1y_2 and y_2 yields z_2y_1.

29. Having solved Problem 28, the rest is easy! First take expressions y_1 and y_2 said that y_1 yields Θ_1y_2 and y_2 yields Θ_2y_1 (as in the last problem, taking Θ_1 instead of z_1 and Θ_2 instead of z_2). Since y_1 yields Θ_1y_2, then Θ_2y_1 yields $\Theta_2(\Theta_1y_2)$. Since y_2 yields Θ_2y_1, then Θ_1y_2 yields $\Theta_1(\Theta_2y_1)$.

Thus we have:

(1) Θ_2y_1 yields $\Theta_2(\Theta_1y_2)$,
(2) Θ_1y_2 yields $\Theta_1(\Theta_2y_1)$.

And so we take x to be Θ_1y_2 and y to be Θ_2y_1 and thus x yields $\Theta_1(y)$ and y yields $\Theta_2(x)$.

We recall that as in the solution to the last problem we take y_1 and y_2 as follows (using Θ_1 and Θ_2 in place of z_1 and z_2).

$y_1 = CQ\Theta_1Q\Theta_2C,$
$y_2 = Q\Theta_2CQ\Theta_1Q\Theta_2C.$

And so our final solution is

$x = \Theta_1Q\Theta_2CQ\Theta_1Q\Theta_2C,$
$y = \Theta_2CQ\Theta_1Q\Theta_2C.$

30. In the solution to the last problem, we simply take R for Θ_1 and also R for Θ_2, and get x = RQRCQRQRC and y = RCQRQRC.

31. An expression that yields its own repeat is RRVLVQRRVLVQ, as the reader can check. How it was found will soon be clear.

32. Yes, it is true, that as with Arnie's system, for any expression y there is an expression x that yields yx. The key point in showing this is to first find some expression,

which I will denote P such that if x yields y, then Px yields yQyQ.

Well, suppose x yields y. Then Vx yields $\overleftarrow{y}$ (the reverse of y). Then LVx yields $Q\overleftarrow{y}$. Then VLVx yields yQ. Then RVLVx yields yQyQ. We thus take P to be RVLV. Thus if x yields y, then Px yields yQyQ. Since this holds for any y, it holds if we take yP instead of y. Thus if x yields yP, then Px yields yPQyPQ. Now, QyPQ yields yP, hence PQyPQ yields yPQyPQ. We thus take x to be PQyPQ, and so x yields yx. Now, P is the expression RVLV, and so for any y, an expression x that yields yx is RVLVQyRVLVQ.

Let us check that RVLVQyRVLVQ really yields yRVLVQyRVLVQ.

Well, QyRVLVQ yields yRVLV.

Hence VQyRVLVQ yields VLVR$\overleftarrow{y}$.

Hence LVQyRVLVQ yields QVLVR$\overleftarrow{y}$.

Hence VLVQyRVLVQ yields yRVLVQ.

Hence RVLVQyRVLVQ yields yRVLVQyRVLVQ.

33. The derivation of the fixed point principle is now the same as with Arnie's system. Given a special expression Θ (any combination of the letters L, V, R), to find a fixed point of Θ, we need an x that yields Θx (as before), and then Θx will yield Θ(Θx), and so Θx will then be a fixed point of Θ. Well, for this system, an x that yields Θx is RVLVQΘRVLVQ, and so a fixed point of Θ is now ΘRVLVQΘRVLVQ

34. A solution is:

x = RVLRVQVQRVLRLVQ

y = VQRVLRLVQYQRVLRLVQQ

How was this found? By the double fixed point principle for this system, which we are approaching.

35. It took me a long time to solve this! The crucial step was my realization that what was needed was an expression Δ such that if x yields y, then Δx yields yQyQQ

(instead of yQyQ, as before). Well, I found solutions by two different methods. My first try was to use the P that we already have (P = RVLV), which has the property that if x yields y, then Px yields yQyQ. Then I reasoned as follows:

Suppose x yields y.

Then Px yields yQyQ.

Then VPx yields $\overleftarrow{Qy}\,\overleftarrow{Qy}$ (the reverse of yQyQ).

Then LVPx yields $QQ\overleftarrow{y}\,Q\overleftarrow{y}$.

Then VLVPx yields yQyQQ.

Thus we take Δ to be VLVP, which is VLVRVLV.

Then I had the uneasy feeling that a simpler solution should be possible: The above Δ involved 4 reversals, and my gut feeling was that only 2 should be necessary, and so I decided to forget P and start from scratch. I asked myself how to get from y to yQyQQ? Well, first reverse y, getting $\overleftarrow{y}$, then put a Q before it, getting $Q\overleftarrow{y}$, then repeat, getting $Q\overleftarrow{y}Q\overleftarrow{y}$, then putting a Q before that, getting $QQ\overleftarrow{y}Q\overleftarrow{y}$, then reversing, getting the desired yQyQQ. Eureka!

And so, suppose x yields y.

Then Vx yields $\overleftarrow{y}$.

Then LVx yields $Q\overleftarrow{y}$.

Then RLVx yields $Q\overleftarrow{y}Q\overleftarrow{y}$.

Then LRLVx yields $QQ\overleftarrow{y}Q\overleftarrow{y}$.

Then VLRLVx yields yQyQQ.

And so we take Δ to be VLRLV. This Δ is of length five, instead of the length seven of the first Δ, and has only two reversals instead of four. And so this simpler Δ is the one we shall take.

Since QyQ yields y, then ΔQyQ yields yQyQQ. Then for any expression w, if we take wΔ for y, we see that ΔQwΔQ yields wΔQwΔQQ, and so z yields wzQ, where z is the expression ΔQwΔQ. Thus we have the following fact:

(1) For any expression w, there is an expression z, namely $\Delta Q w \Delta Q$, such that z yields wzQ.

Now, given special expressions Θ_1 and Θ_2, we seek expressions x and y such that x yields $\Theta_1(y)$ and y yields $\Theta_2(x)$. When we take w to be $\Theta_2 Q \Theta_1$ in fact (1) above, we have the following key fact:

(2) If z is the expression $\Delta Q \Theta_2 Q \Theta_1 \Delta Q$, then z yields $\Theta_2 Q \Theta_1 zQ$.

We thus take z to be $\Delta Q \Theta_2 Q \Theta_1 \Delta Q$. We then take x to be $\Theta_1 z$ and y to be $\Theta_2 Q \Theta_1 zQ$, and we will see that x yields $\Theta_1 y$ and y yields $\Theta_2 x$.

(a) Since z yields y (which is $\Theta_2 Q \Theta_1 zQ$), then $\Theta_1 z$ yields $\Theta_1 y$. Thus x (which is $\Theta_1 z$) yields y.

(b) $Q \Theta_1 zQ$ obviously yields $\Theta_1 z$, and so $Q \Theta_1 zQ$ yields x (which is $\Theta_1 z$), and therefore $\Theta_2 Q \Theta_1 zQ$ must yield $\Theta_2(x)$. Thus y yields $\Theta_2(x)$ (since y is $\Theta_2 Q \Theta_1 zQ$).

This proves that x yields $\Theta_1(y)$ and y yields $\Theta_2(x)$, where $x = \Theta_1 z$ and $y = \Theta_2 Q \Theta_1 zQ$. Since $z = \Delta Q \Theta_2 Q \Theta_1 \Delta Q$, we have the solution:

$x = \Theta_1 \Delta Q \Theta_2 Q \Theta_1 \Delta Q$

$y = \Theta_2 Q \Theta_1 \Delta Q \Theta_2 Q \Theta_1 \Delta QQ$

Since $\Delta = VLRLV$, then in unabbreviated notation, our solution is:

$x = \Theta_1 VLRLVQ \Theta_2 Q \Theta_1 VLRLVQ$

$y = \Theta_2 Q \Theta_1 VLRLVQ \Theta_2 Q \Theta_1 VLRLVQQ$

Much more complicated than Arnie's system!

CHAPTER X
SOME CURIOUS SYSTEMS

§1. Terminating or Non-Terminating?

A while later when I saw Arnie again, he told me he had had a number of visitors after me, who had loved to talk about the kind of machines we enjoyed thinking about. It tool a long time, but he told me in detail about all these visits that day. Let me tell you exactly what Arnie said happened.

The first person who came to visit was a friend of his named Bernie, who told him that after seeing Arnie's machines he had built one that obeyed Arnie's Rules Q and C, as well as other rules.

The first person who had come to visit was a friend of his named Bernie, who told him that after seeing Arnie's machines he had built one that obeyed Arnie's Rules Q and C, as well as other rules.

Bernie described his machine with great enthusiasm as follows:

"The machine accepts certain expressions and not others. Suppose an acceptable expression x is fed into the machine. Then out comes an expression y. If y is not acceptable, that's the end of the process. If y is acceptable, it is fed back into the machine and out comes an expression z. If z is not acceptable, that's the end of the process. If z is acceptable it is fed back into the machine and so the process continues for one more cycle. The process is repeated over and over again, and either it goes on forever, or some unacceptable expression eventually comes out and the process terminates. If the latter, I call x a *terminating* expression. If the former, I will call x non-terminating. An obvious example of a non-terminating expression is CQC, which yields itself. And so the process gives the infinite sequence CQC, CQC, CQC, ..., CQC, ... Other examples of non-terminating expressions are:

"(a) CQQC

"(b) CQQQC

"(c) CQCQ

"(d) CQCQC"

After this description Bernie presented Arnie with the following problem.

Problem 1 (Bernie). Prove that each of these four expressions are non-terminating.

"Now," said Bernie, "I have one problem I am unable to solve. I would like to find some purely mechanical method of determining which expressions are terminating and which are not. I am looking for an expression s—which I would call a *sage*—such that for any expression x, the expression sx is terminating if and only if x is non-terminating. If I had such an expression s, then I could uniformly determine which expressions are terminating and which are not, because I would then build a duplicate of my machine and given any expression x, I would put in x in the original machine and at the same time put sx in the duplicate machine and then start the machines going at the same time. If x is terminating, then the process of the original machine will terminate, and I will know that x is terminating. On the other hand, if x is non-terminating, then sx is terminating, hence the process of the duplicate machines will terminate, and I will then know that x is non-terminating. Thus if I could find a sage, I would then be able to determine which expressions are terminating and which are not."

"And have you found a sage?" asked Arnie with a smile.

"No," replied Bernie. "That's what I have come to you for. With your expertise, surely you can help me find a sage, once I give you all the rules of my machine. As I told you, the first two rules of your first machine are rules of my machine. There are 27 other rules of my machine. The rest of them are—"

"I'm sorry to interrupt you," said Arnie, "but regardless of what your other 27 rules are, and regardless of what other rules you could add, since your machine has Rules Q and C, the existence of a sage is impossible!"

Problem 2. Why is it impossible?

§2. Corey's Problem

Shortly after, another friend named *Corey* paid Arnie a visit and told him of a problem he had. He was working on a project, and told Arnie that to complete the project he needed to design an infinite sequence M_1, M_2, ..., M_n, ... of machines that were to obey certain conditions that he would describe in a moment. The machines would all operate on the positive integers 1, 2, 3, ..., n, so you were to always interpret the word *number* to mean positive integer in the context of Corey's machines. For each of Corey's machines M, you would feed in a number x and out would come a number M(x). This M(x) would be the output of M for the input x. Corey used the notation x* to denote the number $M_x(x)$.

Corey wanted his battery of machines to have the following three properties:

(1) For any machine M there is a machine N such that for every number x, the number N(x) is different from M(x). Such a machine N might be called an *opposer* of M.

(2) For any machine M there is a machine D such that for any number x, the number D(x) is the same as M(x*). Such a machine might be called a *diagonalizer* of M.

(3) There is a machine I (an *identity* machine) such that whatever number you put in, the same number comes out—i.e., for all x the number I(x) is x.

Arnie thought about this for a while, and finally said: "I'm sorry to disappoint you, but there cannot exist a sequence of machines satisfying those conditions."

Problem 3. Show that Arnie was right.

§3. Self-Reproducing Machines

The next time Arnie had visitors, they were four in number, Davey, Ernie, Farleigh and Gordy. All of them were there to talk with Arnie about ambient machines—robots if you will. Each machine of this kind is programmed to construct another machine. If a machine M constructs another machine whose program is the same as that of M, then M is called a *self-reproducing* machine. Each of the visitors had devised a system that could be used to construct a self-reproducing machine. And each friend would leave Arnie a problem as a parting gift.

System 1. The inventor of the first system was Arnie's friend Davey. Arnie said Davey's system used two of Arnie's own rules:

Rule Q. Qx yields x.

Rule R. If x yields y then Rx yields yy.

Each program of his system, Davey had told Arnie, consists of a string of capital letters. He abbreviated "the machine with program x constructs a machine with program y" by "x constructs y" in stating his construction rule:

Rule C. If x yields y then Cx constructs y.

Problem 4 (Davey). Exhibit a program for a self-reproducing machine based on System 1.

System 2. Arnie's friend Ernie had a very different set of yielding rules. And Ernie used a script letter for a variable name to write down his yielding rule, Arnie said, pointing at Rule ℓ on a sheet of paper on his desk. Ernie had told Arnie that for any letter ℓ other than Z, by ℓ' he meant the letter that comes after ℓ, and he took Z' to be A. Thus A' = B; B' = C, ..., Y' = Z and Z' = A. Here is the yielding rule that Ernie gave Arnie after that explanation:

Rule ℓ. For any letter ℓ, and any expression x, the expression ℓ'x yields ℓxx.

Ernie too used Davey's construction rule:

Rule C. If x yields y then Cx constructs y.

Problem 5 (Ernie). Exhibit a program for a self-reproducing machine based on Ernie's System 2.

System 3. After Ernie told Arnie about his system, Farleigh presented his, adopting Ernie's use of a script letter to explain his two construction rules. He defined ℓ and ℓ' as in Ernie's system and his system had the following two construction rules: For any letter ℓ, he defined ℓ' as in the last system. His system had the following two rules:

(1) For any letter ℓ, ℓ' constructs $\ell\,\ell$.

(2) If y constructs Fℓ, then ℓ' constructs y.

Problem 6 (Farleigh). Prove that some machine based on System 3 is self-reproducing.

System 4. Finally Gordy presented a fourth system, his own. Gordy told Arnie that in his system, not every string of capital letters is a program. But for any string x, if xx is a program, then xx is the program of a self-reproducing machine.

Gordy said that his system satisfied the following conditions:

(1) For any expression x, the expression Dxxx is a program.

(2) For any expressions x and y, if xBy is a program, so is xDy.

Problem 7 (Gordy). Find a program for a self-reproducing machine based on System 4.

Solutions to the Problems of Chapter X

1. (a) The expression CQQC yields QCQQC which in turn yields back CQQC. We thus have two expressions, each of which yields the other. So of course they are both non-terminating. You put one in, the other comes out. You put back in the other, and the first one comes out. The process clearly never terminates.

(b) Let us say that an expression x leads to an expression y if by starting the process with x we eventually get y as an output. Let us say the following: x leads to y in one step if x yields y; x leads to y in two steps if x yields some expression that yields y; x leads to y in three steps if x yields some expression which yields some expression which yields y, and so forth. Thus if x yields y and y leads to z in n steps, x leads to y in n + 1 steps. Let us note that if t is any string of Q's, then tx will lead to x in n steps, where n is the number of Q's.

Now, it is obvious that if x leads to itself, then x must be non-terminating. In part (a) of this problem, the expression CQQC led to itself in two steps. We are presently investigating the expression CQQQC. Well, it leads to itself in three steps, since it yields QQCQQQC, which in turn leads to CQQQC in two steps. Hence CQQQC is non-terminating.

In general, if t is any string of Q's, the expression CtC leads to itself, and is therefore non-terminating.

(c) The expression CQCQ is in a way more interesting, in that it does not lead to itself, but is nevertheless non-terminating for a different reason.

Actually I will prove something more general, namely that if t is any string of Q's, then the expression CtCQ must be non-terminating.

Let us call a class of expressions *closed* if every member of the class leads to some member of the class—either itself or some other member. It is obvious that if a class is closed, then all its members must be non-terminating, since no member can lead to an expression outside the class. Well, the class of all expressions of the form CtCQ (t a string of Q's) is closed, because CQCQ yields CQQCQ, which is in the class, and any string of Q's of length 2 or more is of the form Qt, for t a string of Q's, and CQtCQ yields tCQQtCQ, which in turn leads to CQQtCQ, which is a member of the class. Thus each member x leads to another member (which happens to be longer than x) and hence all these expressions are non-terminating.

(d) The expression CQCQC produces CQCQCQC, which in turn produces CQCQCQCQC, and so forth. That is, if M is the expression CQ, or CQ repeated any number of times, the expression CQMC yields MCQMC, which is another expression of the same form (a longer one, in fact), and so all such expressions are non-terminating.

2. Let us first note that if x leads to y, then if y is terminating, so is x, and if y is non-terminating, so is x. Thus if x leads to y, then x is terminating if and only if y is terminating. Now, we recall that in Arnie's machine for any expression y there is an expression x that yields yx. Thus no expression s can be a sage, because there is an x that leads to sx, hence it is *not* the case that sx is terminating if and only if x is *not* terminating (because sx is terminating if and only if x *is* terminating).

3. Fixed points again play a rule here. Call a number x a fixed point of machine M if $M(x) = x$. It follows from condition 2 alone that every machine M has at least one fixed point. Reason: Let M_b be a machine that diagonalizes M. Thus $M_b(x) = M(x^*)$ for every x. Taking b for x,

we have $M_b(b) = M(b^*)$ but $M_b(b)$ is b^*. Thus $b^* = M(b^*)$, which means that b^* is a fixed point of M.

It then follows from conditions (1) and (2) that for any machine M there must be at least one number n such that $M(n) \neq n$, because by condition (1) there is a machine M' such that $M'(x) \neq M(x)$ for every x, and then by condition (2), there is at least on n such that $M'(n) = n$, and hence $M(n) \neq n$. Thus no machine can satisfy condition (3), if conditions (1) and (2) both hold.

4. We obviously need an x that yields Cx. Then Cx will construct Cx.

Such an x is RQCRQ. [Since QCRQ yields CRQ, then RQCRQ yields the repeat of CRQ, which is CRQCRQ]. Since RQCRQ yields CRQCRQ, then CRQCRQ constructs CRQCRQ, and is thus the program of a self-reproducing machine.

5. In Rule 1, take C for ℓ, and we have Dx yields Cxx, for all x. Then take D for x, and we have DD yields CDD. Hence CDD constructs CDD. Thus CDD is the solution.

6. In (2), take F for ℓ and G for y, and we have: (2)': If G constructs FF, then G constructs G. But G *does* construct FF (by (1), taking F for ℓ). Therefore G constructs G.

7. In (2), take DB for x and B for y, and we have: If DBBB is a program, so is DBDB. But DBBB is a program (by (1)), so that DBDB is a program, and hence the program of a self-reproducing machine.

CHAPTER XI

HOW TO STUMP A
DECISION MACHINE

There is a phrase that occurs frequently in mathematics, but which is sometimes confusing to those who haven't seen it before. This phrase is "if and only if," which is actually used quite literally. For example, if I say to a child, "I'll take you to the movies if and only if you are good," I mean that if you are good, I will take you, but I will take you *only* if you are good (in other words, if you are not good, then I won't take you). Thus, given two propositions, if we say that one of them is true *if and only if* the other is true, we mean that if either one is true, so is the other—each one implies the other—in other words that they are either both true or both false. Because the phrase "if and only if" occurs so frequently in mathematics the mathematician Paul Halmos suggested using the abbreviation "iff" for it, and we shall follow him very often in the remainder of this text, especially in solutions where "iff" will abound!

In this chapter the word *number* will mean positive whole number—one of the numbers 1, 2, 3, ..., n, ... And in this chapter, all variables will stand for numbers.

We are given a machine M whose function is to help us determine which numbers have which properties. The machine has infinitely many compartments, or *registers* R_1, R_2, ..., R_n, ... For each n, we call n the *index* of R_n.

Each register is, so to speak, in charge of a certain property of numbers, and its function is to try to determine which numbers have the property and which ones don't.

To determine whether a given number x has a certain property, one first activates the register in charge of the property and then feeds x into that register. The machine then goes into operation and one and only one of three things happens:

(1) The machine eventually halts and flashes a signal (say a green light) signifying that x *does* have the property, in which case we say that the register *affirms* x.

(2) The machine eventually halts and flashes a different signal (say, a red light) to indicate that the number doesn't have the property, in which case we say that the register *denies* x.

(3) The machine runs on forever, unable to decide whether or not x has the property, and thus never halts. In that case we say that x *stumps* the register, or that the register is stumped by x. We call a register *total* iff no number can stump it—in other words, for every number x, the register either affirms x or denies x.

Given two registers R and S and numbers x and y (either different or the same) we say that R behaves towards x as S behaves towards y if either R affirms x and S affirms y, or R denies x and S denies y, or R is stumped by x and S is stumped by y. We say that R and S are *similar* if for every number x, they behave the same way towards x.

We now wish to assign to each number x and each number y (given in that order) a code number which we will denote x∗y, from which we can uniquely recover x and y— that is, for any numbers x, y, z, w, the only way that x∗y can be the same number as z∗w is that x = z and y = w.

There are many ways to do this—for example, we could take x*y to be the number which in ordinary decimal ten notation is written as a string of 1's of length x followed by a string of 0's of length y (in particular, 2*3 would be 11000), but we want our coding to be more general, since several different coding methods will be used in subsequent chapters. All that we now require is that x*y uniquely determine the x and the y. In particular, if x and y are different numbers, then x*x is different from y*y.

We are given that the decision machine M satisfies the following two conditions:

C_1. To each register R is assigned a register $R^\#$ called the *diagonalizer* of R, which behaves the same way towards any number x as R behaves towards x*x. [In general, we will say that a register S *diagonalizes* a register R if for every number x, the register S behaves towards x as R behaves towards x*x. Thus $R^\#$ diagonalizes R.]

C_2. To each register R is assigned a register R' called the *opposer* of R, which affirms those and only those numbers which R denies, and denies those and only those numbers that R affirms (and hence is stumped by those and only those numbers which stump R, if there are any).

A register U will be called *universal* if for any numbers x and y, the register affirms x*y if and only if R_x affirms y.

Problem 1. Prove that if U is a universal register then there is a number that stumps U.

Problem 2. Prove that if a universal register U exists, then there is a register that is stumped by its own index—in other words that there must be a number a such that R_a is stumped by a.

Problem 3. To make matters more concrete, suppose we are given the following three conditions:

(1) The register R_1 is universal.

(2) For any number n, R_{2n} is the opposer of R_n

(3) For any number n, R_{2n+1} is the diagonalizer of R_n.

    (a) Find an odd number n which stumps R_n

    (b) Find an even number n that stumps R_n

    (c) We know that there are two numbers that stump R_1. Are there any others? If so, how many?

    (d) How many universal registers are there?

    (e) How many registers are there that are stumped by their own index?

Problem 4. For any register R, the register $R'^{\#}$ is the diagonalizer of the opposer of R, whereas $R^{\#'}$ is the opposer of the diagonalizer of R. These two registers are not necessarily the same, but are they necessarily similar?

Problem 5. Prove that if one of the registers is universal, then there must exist a register C such that for every number n, the register C affirms n if and only if R_n affirms n.

Problem 6. Prove that for any register R there is a number n such that R behaves towards n*n as R_n behaves towards n.

Problem 7 [A key problem]. Prove that if U is universal, then for any register R there is at least one number a such that U affirms a if and only if R affirms a.

The solution of Problem 1 is an easy consequence of that fact. Why?

Problem 8. Call a register V *contra-universal* if for any numbers x and y, V affirms x*y if and only if R_x *denies* y.

Can a contra-universal register be stumped?

Problem 9. Suppose that W is a register such that for any numbers x and y, the following conditions hold:

    (1) If R_x affirms y then W affirms x*y.

    (2) If R_x denies y, then W doesn't affirm x*y. Can W be stumped?

Problem 10. Suppose that for every x and y, the following conditions hold:

(1) If R_x affirms y, then W affirms x∗y.

(2) If R_x denies y then W denies x∗y. Can W be stumped?

Problem 11 [Halting Problem]. We are saying that a register *halts* at a number n if it either affirms or denies n (and thus does not run on forever). In other words a register halts at n if and only if it is not stumped by n.

It would be nice to have a register H that halts at those and only those numbers x∗y such that R_x does not halt at y, because we would then have a purely mechanical method of determining which numbers stump which registers—namely, to find out whether a given number y stumps a given register R_x, we feed in y to R_x and feed in x∗y to H and set both registers going and wait. Sooner or later either H or R_x will halt, but not both. If R_x halts, we will know that y does not stump R_x.

On the other hand, whenever y is going to stump R_x we will find out about sooner or later that this is the case, since by assumption H must halt in this case when fed in x∗y. As soon as H halts, we will know that R_x can never halt, i.e that y stumps R_x.

Thus it would be nice to have such a register H, but alas, no such register is possible. Why?

Affirmation Sets. For any number n, let A_n be the set of all numbers affirmed by R_n, and B_n be the set of all numbers denied by R_n.

Two sets A and B of numbers are called *disjoint* if they contain no common element. They are called *complementary*—or we say that each of them is the *complement* of the other—if every number x belongs to one and only one of them—in other words if the two sets are disjoint and if no number is outside both of the sets. Thus to say that B is the complement of A is to say that B consists of all and only those numbers that are *not* in A.

Since no register R_n both affirms and denies the same number x, this means that there is no number x that is common to A_n and B_n—in other words, the sets of A_n and B_n are *disjoint*.

By condition C_2, for every register R_n there is a register R_k that affirms just those numbers that R_n denies, and denies just those numbers that R_n affirms. This means that for every n, there is some number k such that $A_k = B_n$ and $B_k = A_n$. Thus in the infinite sequence A_1, A_2, ..., A_n, ... and the infinite sequence B_1, B_2, ..., B_n, ... the *sets* of the two sequences are the same, only arranged in a different order. We shall refer to those sets as the *affirmation* sets. The set A_n is the affirmation set of the register R_n. The set B_n, though not the affirmation set of R_n, is nonetheless an affirmation set—the affirmation set of R_n'.

Problem 12. Suppose that A is a set of numbers and that there is a register R that affirms all numbers in A and denies all numbers not in A. Is A necessarily an affirmation set? Is the complement of A necessarily an affirmation set?

Problem 13. Suppose there is a universal register U. Is the complement of every affirmation set an affirmation set?

Solutions to the Problems of Chapter XI

1. Let U' be the opposer of U and let R_a be the diagonalizer of U'. Then for every number n, the register R_a affirms n if and only if U' affirms n*n, which in turn is the case if and only if U denies n*n. Thus: (1) R_a affirms n iff U denies n*n.

Since U is universal, then also:

(2) R_a affirms n iff U affirms a*n.

Thus for any number n, it follows from (1) and (2) that (3) U affirms a*n iff U denies n*n.

Since (3) is true for *every* number n, then in particular it is true for the number a. This means that U affirms a*a iff U denies a*a, which is impossible, since no register both affirms and denies the same number. Likewise, from (3) it follows that if U denies a*a it also affirms a*a, which again is not possible Thus U neither affirms or denies a*a, and thus is stumped by a*a.

We have now seen that if a is the index of U'# (the diagonalizer of the opposer of U), then a*a stumps U. It is also true that if b is the index of U#' (the opposer of the diagonalizer of U) then b*b also stumps U, because R_b, being the opposer of U#, affirms b if and only if U# denies b, which in turn is the case iff U denies b*b. And so again we have a number b such that R_b affirms b iff U denies b*b (since U is universal), and so U is stumped by b*b.

2. We will see that U'# is stumped by its own index, and also that U#' is stumped by its own index.

Let a be the index of U'#. We have seen in the solution of Problem 1 that a*a stumps U, hence a*a must also stump U' (Why?). Then a must stump U'# (since U'# is the diagonalizer of U'). Thus U'# is stumped by its own index a.

Now let b be the index of U#'. Then, as we have seen in the solution of Problem 1, the number b*b stumps U.

Hence b stumps $U^\#$, and so b also stumps the opposer $U^{\#'}$ of $U^\#$. Thus $U^{\#'}$ is stumped by its own index b.

3. Under the special conditions of this problem, for any number n, the index of R_n' is 2n, hence the index of $R_n'^\#$ is $2(2n) + 1$, which is $4n + 1$. In particular, taking 1 for n, the index of $R_1'^\#$ is 5. Since R_1 is universal (in this special problem) then $R'^\#$ is stumped by its own index (as seen in the solution of the last problem), and so R_5 is stumped by its own index 5. This answers (a).

As seen in the solution to the last problem, for any universal register U, the register $U^{\#'}$ is also stumped by its own index, hence $R_1^{\#'}$ is stumped by its own index, which is 6, since the index of $R_1^\#$ is $(2 \times 1) + 1$, which is 3, hence the index of $R_1^{\#'}$ is 2×3, which is 6. Thus R_6 is stumped by the even number 6, which answers (b).

As for (c), (d), and (e), let us first note the obvious fact that for any register R, the register R'' is similar to R (why?), and also the index of R'' is greater than of the index of R (the index of R'' is 4 times that of R). Now, let U_1 be the universal register R_1, U_2 be U_1'', U_3 be U_2'', etc. (for each number n, $U_{n+1} = U_n''$). All the infinitely many registers U_1, U_2, ..., U_n, ... are universal; they are all distinct and have distinct indexes. Thus there are infinitely many universal registers, and so the answer to (d) is "infinitely many" (more precisely, "denumerably many").

For each number n, let a_n be the index of U_n and b_n the index of $U_n'^\#$, which is $4a_{m \neq 1}$ (as we have seen). If n and m are different numbers, so are a_n and a_m, hence also b_n is distinct from b_m. Thus the registers $U_1'^\#$, $U_2'^\#$, ..., $U_n'^\#$ are all distinct and each register $U_n^\#$ is stumped by b_n, and so there are infinitely many registers that are stumped by their own index, which answers (e).

Finally, the numbers b_1, b_2, ..., b_n, ... are all distinct and for each number n, the number b_n stumps $U_n'^\#$, hence

b_n*b_n stumps U_n (as we have seen in the last problem), and must also stump all the universal registers, since they are all similar to each other. Also, if m is distinct from n, then b_m is distinct from b_n, hence b_m*b_m must be distinct from b_n*b_n (as we recall, our coding is required to satisfy the condition that for any numbers x, y, z, w, if $x*y = z*w$, then $x = z$ and $y = w$, and that if $x \neq y$, then $x*y \neq z*w$). Thus each of the infinitely many numbers b_1*b_1, b_2*b_2, ..., b_n*b_n, ... stumps all the universal registers, hence R_1 in particular, which answers (c).

4. Yes, they are. For any number n, the register $R'^{\#}$ affirms (denies) n iff R' affirms (denies) $n*n$, iff R denies (affirms) $n*n$. Thus $R'^{\#}$ affirms (denies) n iff R denies (affirms) $n*n$.

On the other hand, $R^{\#'}$ affirms (denies) n iff $R^{\#}$ denies (affirms) n iff R denies (affirms) $n*n$. Thus, like $R'^{\#}$, $R^{\#'}$ affirms (denies) n iff R denies (affirms) $n*n$. Thus $R'^{\#}$ affirms (denies) n if and only if $R^{\#'}$ affirms (denies) n, and so the two registers are similar.

5. If U is universal, take C to be $U^{\#}$, and this C does the trick: For any number n, C affirms n iff U affirms $n*n$, which is the case iff R_n affirms n.

6. Given a register R, let n be the index of the diagonalizer $R^{\#}$ of R. Thus for any number x, R_n affirms (denies) x if and only if R affirms (denies) $x*x$. We take n for x, and so R_n affirms (denies) n if and only if R affirms (denies) $n*n$.

7. By the last problem, for any register R there is a number n such that R affirms $n*n$ if and only if R_n affirms n. But also R_n affirms n iff U affirms $n*n$. Therefore R affirms $n*n$ iff U affirms $n*n$, and so take m to be $n*n$, and then R affirms m if and only if U affirms m.

A solution to Problem 1 is immediate from this. Take R to be U', and so there is a number m such that U affirms m iff U' affirms m—thus U affirms m if and only if U denies m. Since no register can both affirm and deny the same number, then U neither affirms or denies m—it is stumped by m.

8. Of course it can. To begin with, for any register R there must be a number m such that R denies m iff V affirms m. [Reason: By Problem 6, there is a number n such that R denies n*n iff R_n denies n, but R_n denies n iff V affirms n*n, and so R denies n*n iff V affirms n*n. So take m to be n*n].

Now take R to be V itself (instead of U', as in the last problem). Thus there is a number m such that V denies m iff V affirms m, and so V is stumped by m.

9. We are given:

(1) If R_x affirms y, then W affirms x*y.

(2) If R_x denies y, then W does not affirm x*y.

We now consider the opposer W' of W. By Problem 6, there is a number n such that R_n behaves towards n as W' behaves towards n*n. Hence: (1)' R_n affirms n iff W denies n*n (since W' affirms n*n iff W denies n*n).

(2)' R_n denies n iff W affirms n*n. Now, if W affirmed n*n, then by (2)', R_n would deny n, hence by (2), taking n for both x and y, W would not affirm n*n, which is logically impossible. Hence W does not affirm n*n.

If W denied n*n, then by (1)', R_n would affirm n, and then by (1) W would affirm n*n, which is not possible since no register affirms and denies the same number. Thus n*n stumps W.

10. Of course it can! This is immediate from the last problem, because if W_x denies x*y, it certainly does not affirm x*y, and thus the given conditions of this problem imply the given conditions of the last problem.

11. This is immediate from Problem 6. Given a register H, by Problem 6 there is at least one number n such that

H behaves towards n*n as R_n behaves towards n. Thus it cannot be that H halts at n*n if and only if R_n does not halt at n. Thus there are numbers x and y—namely x = n and y = n—such that it is not the case that H halts at x*y iff R_x is stumped by y.

12. Suppose that R affirms all numbers in A and denies all numbers that are not in A. Now, R cannot affirm any number not in A, since R would deny any such number. Thus all numbers affirmed by R are in A. Thus R affirms all and only those numbers that are in A, hence A is the affirmation set of R. Thus A is indeed an affirmation set.

Let $\overline{A}$ be the complement of A. Since R affirms all numbers in A and denies all numbers in $\overline{A}$ then R' denies all numbers in A and affirms all numbers in $\overline{A}$. Thus R' is a register that affirms all numbers in $\overline{A}$ and denies all numbers that are not in $\overline{A}$ (since they are in A). Thus $\overline{A}$ is also an affirmation set.

13. The answer is *no*, because we will see that the affirmation of U—call it A—is an example of a set whose complement is *not* an affirmation set!

Let B be any affirmation set. We will show that B cannot be the complement of A.

B is the affirmation set of some register R. R affirms all and only those numbers that are in B. By Problem 7 there is at least one number n such that U affirms n if and only if R affirms n. Thus n is in A if and only if n is in B. This means that either n is in both A and B, or in neither A nor B, and so B cannot be the complement of A. Thus no affirmation set is in the complement of A and so the complement of A is not an affirmation set.

CHAPTER XII

SOME ADDITIONAL
GÖDELIAN PUZZLES

Consider the following sentence:

THIS SENTENCE CAN NEVER BE PROVED.

Assuming that everything provable is true, we get the following paradox: If the above sentence were false, then what it says would not be the case, which would mean that it *can* be proved, contrary to the fact that only true sentences can be proved. Therefore it is contradictory to assume the sentence is false; it must be true. I have now proved that the sentence is true. Since it is true, then what it says is the case, which means that it can never be proved, so how come I just proved it?

Problem 1. How do you get out of that paradox?

Now let's consider a variant: Suppose we have a system—call it System S—that proves various English sentences, and suppose the system is correct, in that it proves only true sentences. Consider now the following sentence:

THIS SENTENCE IS NOT PROVABLE IN SYSTEM S

Problem 2. Does the above sentence lead to a paradox? What is the status of that sentence? Is it true or false? Is it provable in system S or not, or is there no way of telling?

The above little puzzle embodies an essential idea in the proof of Gödel's famous theorem known as the

Incompleteness Theorem [Gödel, 1931]: At the turn of the 20th century there appeared two mathematical systems so comprehensive that it was taken for granted that every true mathematical statement, stated in the language of the systems, could be proved in them, if one could only discover the proof. Well, in 1931 Kurt Gödel startled the entire mathematical world by showing that for each of those two systems, as well as for a variety of related systems, there were true sentences not provable in the system. What Gödel did was assign to each sentence of the system a number, subsequently called the "Gödel number of the sentence," and then by a very ingenious method (using a fixed-point principle) he constructed a sentence X that asserted that a certain number n was the Gödel number of a sentence that was not provable in the system, but the amazing thing is that n was the Gödel number of the very sentence X itself! Thus X asserted that its own Gödel number was the number of an unprovable sentence—in other words, X asserted its own non-provability in the system. This means that X is true if and only if it is not provable in the system—in other words, either the sentence is true but not provable (in the system), or false but provable. The latter alternative is ruled out by the fact that the system is correct in the sense that only true sentences are provable in the system. Thus the sentence X is true but not provable in the system. [This sentence has been paraphrased "I am not provable."]

How did Gödel construct such an amazing sentence? That is one of the main topics of this book. In this chapter I will present some puzzles inspired by and related to Gödel's theorem. In the next chapter I will first construct some miniature systems that embody Gödel's idea in a very instructive way, and then state certain conditions whose fulfillment allows Gödel's argument to go through.

Problem 3. Suppose I make you the following offer: I place a quarter and a penny on the table and ask you to make a statement. If the statement is true, I will give you one of the two coins, not saying which one. If the statement is false, I give you neither coin. What statement could you make such that my only option would be to give you the quarter (assuming I kept my word). [How this is related to Gödel's theorem is discussed in the solution.]

Problem 4. As a variant, there is another statement you could make which would force me to give you both coins! What statement would work?

Problem 5. There is still another statement you could make which would make it impossible that I keep my word. What statement could that be?

Problem 6. For years I have been using the problem of the penny and quarter with my logic students, until one day I realized to my horror that the student could make a statement such that the only way I could keep my word is by paying him a million dollars! What statement would work?

Problem 7. Here is a problem which should bear an obvious resemblance to Gödel's theorem: A logician once visited the Island of Knights and Knaves, where knights make only true statements and knaves make only false statements, and each inhabitant of the island is either a knight or a knave. The logician was a hundred percent accurate in his proofs—anything he could prove was really true. The logician came across a native who made a statement from which it follows that the native must be a knight but the logician can never prove that he is. What statement could work?

Problem 8. In the solution to the last problem we have proved that the native is a knight, and also that the logician can never prove that he is. Now, the logician knows logic as well as you and I, and so what prevents him or her to give the same proof that the native is a knight? Is there something you and I know which the logician doesn't?

Solutions to the Problems of Chapter XII

1. The fact is that the notion of *provability* is not well defined. Given a mathematical system with axioms and rules for deriving various sentences from the axioms, the notion of provability *for that system* is well defined, but as yet, no one has come up with the notion of provability in the absolute sense, independent of a given system.

2. Now there is no paradox, but rather an interesting truth—namely that the sentence must be true but not provable in System S; because if the sentence were false, then contrary to what it says, the sentence *would* be provable in System S, but it is given that only true sentences are provable in System S. Therefore the sentence cannot be false; it must be true. Since it is true, then as it says, it is not provable in System S. Thus the sentence is true, but not provable in System S.

3. A statement that works is "You will not give me the penny." If the statement were false, then contrary to what it says, I *would* give you the penny, but I cannot give you either coin for a false statement. Therefore the statement must be true, and therefore, as it says, I will not give you the penny. But I must give you one of the two coins for a true statement, hence I have no option but to give you the quarter.

In relation to Gödel's theorem, I was thinking of the penny as corresponding to provability and the quarter as corresponding to truth. And so the statement "You will not give me the penny" is the analogue of Gödel's sentence, which can be thought of as saying "I am not provable"

4. A statement that works is "You will give me either both coins or neither coin."

The only way the statement could be false is that I give you exactly one of the coins, but I cannot give you a coin for a false statement. Therefore the statement must be true, which means that I either give you both coins or give you neither coin. But if I give you neither coin, I am violating the agreement that I must give you one of the coins for a true statement. Therefore my only option is to give you both coins.

5. Obviously a statement that works is "You will give me neither coin." If I give you one of the coins, that makes the statement false, and I have thus violated the condition that I don't give you one of the coins for a false statement. On the other hand, if I don't give you either coin, that makes the statement true, and I have thus failed to give you a coin for a true statement, which again violates my agreement!

6. A statement that works is "You will give me neither the penny nor the quarter, nor a million dollars."

If the statement was true, then, as it said, I would give you neither the penny nor the quarter nor a million dollars, which is contrary to the fact that I must give you either the penny or the quarter for a true statement. Hence (for me to keep my word) the statement cannot be true. Since it is false, then contrary to what it says, I must give you either the penny or the quarter or a million dollars, but I may not give you either the penny nor the quarter for a false statement, hence I have no option but to give you a million dollars.

7. There are several statements that could work, and a particularly simple one, and one which comes closest to Gödel's famous sentence which asserts its own non-provability, is that the native said to the logician, "You can never prove that I am a knight." If the native were a knave, his statement would be false, which would mean that the logician could prove that the native was a knight, hence prove something wrong, which is contrary

to the given fact that the logician could only prove correct facts. Therefore the native can't be a knave; he must be a knight. Since he is a knight, his statement must be true, which means that the logician can never prove that he is a knight. Thus the native is really a knight, but the logician can never prove that he is.

8. Is there something the logician doesn't know which you and I know? Yes, there is! I told you that the logician was completely accurate in his proofs; I never told you that the logician *knew* he was completely accurate! Indeed if the logician believes he was completely accurate he could easily become inaccurate, in fact inconsistent! Suppose the logician believes that everything he could prove was true. Then he could reason thus: "Suppose he (the native) is a knave. Then his statement is false, which means that I *could* prove he is a knight. But my proving him a knight means he really is a knight." [Here the logician is appealing to his own accuracy!]. "Thus if he is a knave, he must also be a knight, which is impossible. Therefore he is a knight. He said that I could not prove he is a knight, but I just did! Hence he was wrong, and so he is a knave." At this point the logician has become inconsistent!

PART II

PROVABILITY, TRUTH AND THE UNDECIDABLE

CHAPTER XIII
TRUTH AND PROVABILITY

As already mentioned, Gödel showed that for mathematical systems of sufficient strength, there must always be sentences that are neither provable nor disprovable in the system [Gödel, 1931]. Closely related to this is a result of the logician Alfred Tarski [Tarski, 1932] that for the type of systems subject to Gödel's argument, truth of the sentences of the system cannot be defined in the system (in a sense that I will later make more precise). In this chapter we consider some very simple systems which, despite their simplicity, illustrate some of the essential ideas behind the Tarski and Gödel results.

§1 Re Tarski

We consider a set of elements called *expressions*. Some expressions are classified as *sentences* and some as *predicates*. Each predicate is the name of a set of expressions. A set of expressions will be called *nameable* if some predicate names it.

To each expression X and each expression Y is associated an expression denoted XY such that if X is a predicate then XY is a sentence. For any predicate H and any expression X, the sentence HX is interpreted to mean that X is a member of the set named by H, and we accordingly define the sentence HX to be *true* if X really does belong to the set named by H.

We are given that the system satisfies the following two conditions:

C_1. To each predicate H is assigned a predicate denoted $H^\#$ called the diagonalizer of H, such that for any expression X, the sentence $H^\#X$ is true if and only if the sentence H(XX) is true. [By H(XX) is of course meant the sentence HY, where Y is the expression XX.]

C_2. To each predicate H is assigned a predicate denoted H' such that for every expression X, the sentence H'X is true if and only if the sentence HX is not true.

From just these two conditions, something surprising follows (at least I hope the reader will be surprised!), namely:

Theorem T [After Tarski, 1932].

The set of true sentences is not nameable.

Before proving Theorem T, we will first state and prove something else, which not only gives particular insight into Theorem T, but is of interest in its own right.

We shall call two sentences *semantically equivalent* to mean that if either one is true, so is the other—in other words, they are either both true or both false (not true). In this chapter, "equivalent" will mean *semantically equivalent*, as opposed to a different kind of equivalence considered in the next chapter.

For any predicate H, by a *semantic fixed point* of H we shall mean a sentence X such that X is equivalent to HX. In this chapter, *fixed point* will mean *semantic* fixed point, as opposed to another meaning of *fixed point* starting in the next chapter.

Note: The notion of *fixed point* is closely related to the subject of self-reference. A fixed point X of a predicate H is a sentence which is true if and only if HX is true, which means that X is true if and only if X belongs to the set named by H. We can informally think of a fixed point X of H as asserting, "I belong to the set named by H."

Now here is the theorem we will use as preparation for the proof of Theorem T.

Theorem F_0 [Semantic Fixed Point Theorem].

Every predicate H has a fixed point.

Problem 1. Prove Theorem F_0.

Now for Theorem T: We will call a predicate H a *truth predicate* if it is the name of the set of all true sentences. Thus Theorem T is to the effect that no predicate is a truth predicate. Now, if H is a truth predicate then for every sentence X, the sentence X if true if and only if HX is true—on other words every sentence X is equivalent to HX, which means that every sentence X is a fixed point of H! Thus to show that H is *not* a truth predicate, it suffices to show that there is at least one sentence X that is *not* a fixed point of H. Such a sentence X might aptly be called a witness that H is not a truth predicate.

Problem 2. Prove Theorem T by showing that for any predicate H there is witness that H is not a truth predicate.

In fact, given a predicate H, one can explicitly write down a sentence using the three symbols H, ' and $^{\#}$ which is a witness that H is not a truth predicate.

Exhibit such a sentence.

Problem 3. We have seen that for any predicate H, a fixed point of H' is H'$^{\#}$H'$^{\#}$. However H$^{\#}$'H$^{\#}$' is also a fixed point of H'. Prove this.

Call two predicates H and K *semantically similar* if for every expression X, the sentences HX and KX are semantically equivalent.

Problem 4. Given a predicate H, consider the two predicates H'$^{\#}$ and H$^{\#}$'. In general they are not the same. Prove that they are nevertheless similar.

Problem 5. We have seen two fixed points H'#H'# and H#'H#' of H', both of length 6. There are two others of length 6, also using just the three symbols H, ', #. Can you find them?

§2 Enter Gödel

We now turn to a system whose predicates and sentences are the same as before and which obeys conditions C_1 and C_2. But now, certain sentences are classified as *provable*. The system is wholly accurate in that only true sentences are provable. In this system the set of provable sentences *is* nameable—its name is the symbol "P."

Theorem GT [After Gödel, with shades of Tarski]. There is a true sentence that is not provable in the system.

Problem 6. Prove Theorem GT. Actually such a sentence can be exhibited using the three symbols P, ' and #. Write down such a sentence.

Refutable Sentences. In addition to the system having certain provable sentences, there is a procedure for *disproving* certain sentences. Such sentences are called *refutable* sentences. The set of refutable sentences is also nameable—its name is the symbol "R." Also in the system, for any predicate H and any expression X, H'X is provable if and only if HX is refutable.

The system is also accurate in that no true sentence is refutable—only false sentences are refutable.

Problem 7 [After Smullyan]. Using any of the three symbols R, ', #:

(a) Write down a sentence that is false but not refutable.

(b) Write down a sentence that is true but not provable.

Remarks. The sentence of Problem 6 is essentially Gödel's famous sentence, which can be thought of as asserting its own non-provability. It is as if it said, "I am not provable." It is like the native of Problem 1 of Chapter

V who said "I am not a certified knight." He is indeed a knight but not a certified one. It is Theorem GT that made me think up that problem. I thought of knights as playing the role of true sentences, certified knights as provable sentences, knaves as false sentences and certified knaves as refutable sentences.

As for my dual version in Problem 7, the sentence of Part (a), which is false but not refutable, can be thought of as saying, "I am refutable." It is false but not refutable. It is like the native of Problem 2 of Chapter V, who said, "I am an uncertified knight." Now if the native were in fact a knave, he would be saying something false (since knaves always lie). But it would be impossible to refute his statement, because we saw in Chapter V that, giving what he said, all the listener could deduce was that he was one of the following: (1) a certified knave; (2) an uncertified knave; (3) an uncertified knight. And there's no way to know which of the three he actually is.

Relation to Standard Systems. What we have done so far, and what we will continue to do in the next two chapters, is to abstract the essential elements of the Tarski and Gödel results. We have so far regarded predicates as names of expressions, whereas in the standard systems investigated by Tarski and Gödel, predicates (more commonly called *formulas*) are names of sets, not of expressions, but of non-linguistic entities such as numbers. In particular, for numerical systems that we will consider later, predicates can be thought of as properties or sets of the natural numbers, 0, 1, 2, ..., n, ... To each H and each natural number n is associated a sentence denoted H(n) (or more accurately H($\bar{n}$), where $\bar{n}$ is the name of the number n), and the sentence H(n) is to be thought of as expressing the proposition that n does belong to the set named by H. For technical reasons, some authors (including the present one) have found it convenient to assign to *every* expression X, whether a predicate or not, and each natural number n the expression $X(\bar{n})$ (sometimes written $X < \bar{n} >$), not necessarily a sentence, such that if X is a predicate, then $X(\bar{n})$ is a sentence.

Now, Gödel assigned to each expression X a natural number, subsequently called the *Gödel number* of X. What I have done here is to assign to each expression X and each expression Y an expression XY such that if X is a predicate, XY is a sentence. To apply this to standard numerical systems, simply take XY to be the expression $X(\bar{n})$ where n is the *Gödel number* of Y. In this manner we can avoid constant reference to the Gödel numbering.

Solutions to the Problems of Chapter XIII

1. Given a predicate H, for any expression X, the sentence $H^\#X$ is equivalent to H(XX). Since this is so for *every* expression X, it is also so for the expression $H^\#$. Thus in the statement "$H^\#X$ is equivalent to H(XX)" we replace every occurrence of "X" by "$H^\#$," and we see that $H^\#H^\#$ is equivalent to H($H^\#H^\#$). Thus $H^\#H^\#$ is a fixed point of H. [Clever, huh? Thanks Gödel!]

2. We are to show that for any predicate H, there is a sentence X that is not a fixed point of H. Well, obviously no fixed point of H' can also be a fixed point of H, and we have just seen that H', like any other predicate, has a fixed point—namely $H'^\#H'^\#$. Thus $H'^\#H'^\#$ is a witness that H is not a truth predicate. This proves Theorem T.

3. For any predicate H and expression X, the sentence $H^{\#\prime}X$ is true iff $H^\#X$ is not true, which in turn is the case iff H(XX) is not true. Thus:

(1) $H^{\#\prime}X$ is true if and only if H(XX) is not true. Also H(XX) is not true iff H'(XX) is true. From this fact, together with (1) we see that $H^{\#\prime}X$ is true iff H'(XX is true). Thus:

(2) $H^{\#\prime}X$ is equivalent to H'(XX).

In (2) we take $H^{\#\prime}$ for X, with the result that we see that $H^{\#\prime}H^{\#\prime}$ is equivalent to H'($H^{\#\prime}H^{\#\prime}$). Thus $H^{\#\prime}H^{\#\prime}$ is a fixed point of H'.

4. Given a predicate H and any expression X, we have seen in the solution of the last problem (Fact (2)) that $H^{\#\prime}$ is equivalent to H'(XX). But also $H'^\#X$ is equivalent to H'(XX) (by condition C_1 applied to H' instead of H). Hence the sentences $H^{\#\prime}X$ and $H'^\#X$, both being equivalent to H'(XX) are equivalent to each other. Thus $H^{\#\prime}$ is similar to $H'^\#$

5. We know that $H'^{\#}H'^{\#}$ is a fixed point of H'. Also, by the last problem, $H^{\#\prime}$ is similar to $H'^{\#}$, hence $H^{\#\prime}H'^{\#}$ is equivalent to $H'^{\#}H'^{\#}$ and is therefore also a fixed point of H'. By Problem 3, $H^{\#\prime}H^{\#\prime}$ is another fixed point of H', and since $H'^{\#}$ is similar to $H^{\#\prime}$, then $H'^{\#}H^{\#\prime}$ is a fixed point of H'. Thus the four sentences $H'^{\#}H'^{\#}$, $H^{\#\prime}H'^{\#}$, $H^{\#\prime}H^{\#\prime}$ and $H'^{\#}H^{\#\prime}$ are all fixed points of H'.

6. Since the set of provable sentences is nameable and the set of true sentences is not (Theorem T), it follows that the two sets cannot be the same. Thus either some provable sentence is not true, or some true sentence is not provable. The former alternative is ruled out by the given fact that only true sentences are provable, hence the latter alternative must be the case—some true sentence is not provable. How to find such a sentence? Well, any fixed point of P' will do the trick, for suppose X is a fixed point of P'. Then X is true iff $P'X$ is true, and $P'X$ is true iff PX is not true, and PX is not true iff X is not provable (since PX is true iff X *is* provable, because P is the name of the set of provable sentences). Thus X is true iff X is not provable, and since only true sentences are provable, X must be true but not provable.

We know that $P'^{\#}P'^{\#}$ is a fixed point of P', and so $P'^{\#}P'^{\#}$ is a true sentence that is not provable in the system.

7. (a) If X is any fixed point of R, then X is false but not refutable. *Reason*: Suppose X is a fixed point of R. Then X is true iff RX is true, which is the case iff X is refutable (since R is the name of the set of refutable sentences). Thus X is true iff X is refutable. This means that either X is both true and refutable, or neither true nor refutable. The former alternative is ruled out by the given condition that no true sentences are refutable—only false sentences are. Thus the latter alternative holds—X is neither true nor refutable.

By the solution to problem 1, we know that a fixed point of R is $R^{\#}R^{\#}$. Thus $R^{\#}R^{\#}$ is a sentence that is neither true nor refutable.

(b) We are given that in this system, for any predicate H and expression X, the sentence H'X is provable iff HX is refutable. Taking $R^{\#}$ for H and $R^{\#}$ for X, the sentence $R^{\#}{}'R^{\#}$ is provable iff $R^{\#}R^{\#}$ is refutable. Since $R^{\#}R^{\#}$ is not refutable (as we have seen), $R^{\#}{}'R^{\#}$ is not provable. Thus $R^{\#}{}'R^{\#}$ is a sentence that is true but not provable.

CHAPTER XIV

SYNTACTIC INCOMPLETENESS THEOREMS

Semantics is concerned with such notions as truth and meaning. Syntax deals only with combinatorial properties of expressions. In the last chapter the incompleteness theorem GT that we gave appealed to the notion of truth. This is a departure from what Gödel did, for the notion of truth was precisely defined only later by Alfred Tarski. Gödel spoke only about provability, which is a purely syntactic notion. To know whether a purported proof is really a proof, one does not even have to understand what the sentences mean—one need only verify that it was constructed according to the rules of the system, which are purely combinatorial and make no reference to meaning. Indeed one can easily program a computer to check whether a purported proof is really a proof. As I mentioned earlier, the late computer scientist Professor Soul Gorn humorously defined a *formalist* as one who cannot understand a system unless it is meaningless.

In this chapter we consider a purely syntactic generalization of Gödel's theorem with several variants and some related results of the logician J. B. Rosser.

The General Setup. Again we consider a set of elements called *expressions*, some of which are classified as *sentences*, some as *predicates* and to each expression X and each expression Y is assigned an expression denoted XY such that if X is a predicate, then XY is a sentence. Some sentences are classified as *provable* and some as *refutable*. A sentence is called *decidable* if it is either provable or refutable, and *undecidable* if it is neither provable nor refutable. This chapter is devoted to the study of sufficient conditions for the existence of undecidable sentences.

To each expression X is assigned an expression denoted $\overline{X}$ or X' called the *negation* of X, such that if X is a sentence, so is its negation, and if X is a predicate, so is its negation. For a sentence X, I will usually write $\overline{X}$ for the negation of X, but for a predicate H, it will be more convenient to use H' for the negation of H.

We are given that for any sentence X, the sentence X is refutable iff its negation $\overline{X}$ is provable, and X is provable iff $\overline{X}$ is refutable.

We will call two sentences *syntactically equivalent* to mean that if either one is provable, so is the other, and if either one is refutable, so is the other—in other words, if they are either both provable, both refutable, or both undecidable. In this chapter and the next, "equivalent" will mean syntactically equivalent, since semantics now has no place (except in informal discussions).

For the present system we are given the following two purely syntactical conditions:

C_1. For any predicate H and any expression X, the sentence H'X is the negation of HX [Thus $H'X = \overline{HX}$ and so H'X is provable (refutable) iff HX is respectively refutable (provable).

C_2. To each predicate H is assigned a predicate $H^\#$, again called the diagonalizer of H, such that for every expression X, the sentence H#X is (syntactically) equivalent to H(XX) (which again means HXX as it did in the last chapter).

So much will be seen to follow from just those two conditions!

A predicate H will be called a *provability predicate* if for every expression X, the sentence HX is provable iff X is a provable sentence. [This implies that if X is a sentence, then X is provable iff HX is provable].

The system is called *consistent* if no sentence is both provable and refutable, and *inconsistent* if some sentence is both provable and refutable. We assume the system obeys the following classical rule of logic, that if a system is inconsistent then it breaks down in the sense that *all* sentences become provable. In all that follows we will assume that the system is consistent.

Theorem G_0 [After Gödel].

(a) If the system has a provability predicate then some sentence is undecidable.

(b) More specifically if H is a provability predicate then there is an expression X such that HX is undecidable.

Problem 1. Prove Theorem G_0 by exhibiting an undecidable sentence of the form HX, using the three symbols H, ', and # (where H is a provability predicate).

Diagonal Sentences. By a *diagonal* sentence I shall mean a sentence of the form HH for some predicate H.

Problem 2. Prove the following variant of Theorem G_0.

Theorem G_1. If the system has a provability predicate then some diagonal sentence is undecidable.

Representability. We shall call a set W of expressions *representable* if there is a predicate H such that W is the set of all expressions X such that HX is provable, and such a predicate H will be said to represent W. Thus to say that

H represents W is to say that for all expressions X, the sentence HX is provable if and only if X is in W.

For example, a provability predicate is a predicate that represents the set of all provable sentences.

By the *complement* of a set W of expressions, denoted $\widetilde{W}$, is meant the set of all expressions that are not in W.

Theorem C [C for "complementation"]. If some set is representable whose complement is not representable, then some sentence is undecidable.

More specifically, if H represents a set whose complement is not representable then HX is undecidable for some expression X.

Problem 3. Prove Theorem C.

Refutability Predicates. We shall call a predicate K a *refutability* predicate if for all expressions X, the sentence KX is provable iff X is refutable—in other words, if K represents the set of all refutable sentences.

Theorem S_0 [After Smullyan]. If K is a refutability predicate then KX is undecidable for some expression X.

Problem 4. Prove Theorem S_0 by exhibiting such a sentence KX using any of the symbols K, ', #.

Theorem S_0 also has the variant:

Theorem S_1. If the system has a refutability predicate then some diagonal sentence is undecidable.

Problem 5. Assuming K to be a refutability predicate, exhibit a undecidable diagonal sentence, using the same symbols as in Problem 4.

In the last chapter we defined two sentences to be semantically equivalent if they are either both true or both not true. We now define two sentences to by *syntactically* equivalent to mean that if either one is provable, so is the other, and if either one is refutable, so is the other—in other words, they are either both provable, both refutable, or both undecidable. Until further notice (which won't be for several subsequent chapters) *equivalent* will mean syntactically equivalent.

Creative Predicates. We shall call a predicate K *creative* if for every predicate H there is at least one sentence X such that KX is provable iff HX if provable.

Theorem Cr_1. If K is a creative predicate then there is at least one sentence X such that KX is undecidable.

Problem 6. Prove Theorem C_1.

Theorem Cr_2. (a) Any provability predicate is creative (b) Any refutability predicate is creative.

Problem 7. Prove Theorem C_2.

Syntactic Fixed Points. We shall call a sentence X a *syntactic* fixed point of a predicate H if X is (syntactically) equivalent to HX.

Theorem F [Syntactic Fixed Point Theorem].

Every predicate has a syntactic fixed point—moreover one which is a diagonal sentence.

Problem 8. Prove Theorem F. [The proof is virtually the same as the proof of the semantic fixed point theorem F_0 of the last chapter, replacing "semantical equivalence" by "syntactic equivalence."]

Until further notice, "fixed point" will mean syntactic fixed point.

Discussion. Had I proved Theorem F earlier, I could have given much simpler proofs of several of the earlier theorems. The reason I didn't was that I thought it best to first solve the problems the hard way, to give the reader a better appreciation of the value of the fixed point theorem. The use of fixed points not only simplifies proofs of various theorems, but often gives sharper results in the sense of providing more information than is given by the statement of the theorem. The reader will soon see this. As a first example, here is a sharpening of Theorem Cr_2:

Theorem $Cr_{2.1}$. (a) If H is a provability predicate then for any predicate K, the sentence HX is provable iff KX is provable, where X is any fixed point of K. (b) If H is a refutability predicate, then for any predicate K, the sentence HX is provable iff KX is provable, where X is any fixed point of K'.

Problem 9. Prove theorem $Cr_{2.1}$. [This is relatively easy!]

Note: You see how much easier it was to prove Theorem $Cr_{2.1}$ than Theorem Cr_2. Moreover, the statement of Theorem $Cr_{2.1}$ gives more information about the nature of the X in question. Actually in the proof I gave of Theorem Cr_2, fixed points were constructed in the course of the proof, though they were not yet labelled as such.

Next, here is a theorem that simultaneously proves Theorems G_0, G_1, S_0, and S_1—in fact proves sharper versions of them.

Theorem A. (a) If H is a provability predicate, then any fixed point X of H' is undecidable, and so is HX.

(b) If K is a refutability predicate, then any fixed point X of K is undecidable, and so is KX.

Note: Obviously (a) together with Theorem F provides an alternative proof of Theorem G, but it also provides a proof of Theorem G_1, by taking X to be a *diagonal* fixed point of H'. Similarly with (b) and Theorem S and S_1.

Problem 10. Prove Theorem A. [Again, quite easy!]

Discussion. I hope the reader now sees the value of the fixed point theorem. Gödel did indeed construct fixed points in the course of his proof, but never labelled them as such. I believe it was the philosopher Rudolph Carnap who first explicitly stated the fixed point theorem.

The following theorem is particularly useful:

Theorem B. For any representable set W of expressions there are diagonal sentences X and Y such that

(a) X is provable iff X is in W.

(b) Y is refutable iff Y is in W.

Problem 11. (1) Prove Theorem B.

(2) Theorem B provides about the quickest imaginable proofs of Theorems G_1 and S_1. Why?

In all that follows we let P be the set of provable sentences and R the set of refutable sentences.

Problem 12. Which, if either, of the following statements are true?

(a) The complement $\tilde{P}$ of P is not representable.

(b) The complement $\tilde{R}$ of R is not representable.

Note: The answer to the above problem, together with Theorem C, provides another proof of Theorem G_0 and S_0.

Skew Diagonal Sentences. By a *skew diagonal sentence* we shall mean a sentence of the form KK', for some predicate K.

The following is a curious but useful variant of the Fixed Point Theorem.

Theorem F_1. For any predicate H there is a skew diagonal sentence X which is equivalent to $H\overline{X}$. [We recall that $\overline{X}$ is the negation of X].

This theorem has the following corollary

Corollary F_1. For any representable set W there are skew diagonal sentences X and Y such that

(a) X is provable iff $\overline{X}$ is in W.

(b) Y is refutable iff $\overline{Y}$ is in W.

Problem 13. Prove Theorem F_1 and its corollary. The following are further variants of Theorems G and S:

Theorem G_2. If there is a provability predicate then some skew diagonal sentence is undecidable.

Theorem S_2. If there is a refutability predicate then some skew diagonal sentence is undecidable.

Problem 14. Prove Theorems G_2 and S_2. [They can be proved from scratch, but far more easily using Corollary

F_1. It would be a good exercise to try first proving them from scratch.]

Problem 15. Which, if either, of the following statements are true?

(a) If some diagonal sentence is undecidable then some skew diagonal sentence is undecidable.

(b) If some skew diagonal sentence is undecidable then some diagonal sentence is undecidable.

Another Approach. For any set W of expressions we define W* to be the set of all expressions X such that XX is in W.

Theorem W. If W* is representable then there are skew diagonal sentences X and Y such that

(a) X is provable iff $\overline{X}$ is in W.

(b) Y is refutable iff $\overline{Y}$ is in W.

Problem 16. Prove Theorem W.

In what follows we recall that P is the set of provable sentences and R the set of refutable sentences. Thus P* is the set of all X such that XX is a provable sentence, and R* is the set of all X such that XX is a refutable sentence.

Here are two more variants of Theorems G_0 and S_0 which have all four theorems G_1, S_1, G_2, S_2 as corollaries.

Theorem G_0^*. If P* is representable then there is both an undecidable diagonal sentence and an undecidable skew diagonal sentence.

Theorem S_0^*. If R* is representable then there is both an undecidable diagonal sentence and an undecidable skew diagonal sentence.

Problem 17. Prove Theorems G_0^* and S_0^*.

Problem 18. Which, if either of the following two statements are true?

(a) If W is representable so is W*.

(b) If W* is representable so is W.

Problem 19. We stated that Theorems G_1, S_1, G_2, and S_2 are corollaries of Theorems G_0^* and S_0^*. Why is that so? Also, Corollary F_1 is a consequence of Theorem W. Why?

Enter J. B. Rosser. Gödel proved his incompleteness theorem by constructing a predicate that represents P*, but to show that it did represent P*, he had to assume that the system had a stronger property than consistency—a property now known as *omega-consistency*, something we will study in the next chapter. A few years later, the logician J. Barkley Rosser [Rosser, 1936] discovered a more complex predicate, that did not necessarily represent P*, but nevertheless yielded an undecidable sentence, without having to assume Gödel's stronger hypothesis of omega consistency. What now follows is based on Rosser's important contribution.

Separability. Given any two sets A and B, A is said to be a *subset* of B, or B a *superset* of A, if all elements of A are also elements of B—in other words, if B contains all elements of A and possibly other elements as well. Two sets are called *disjoint* if they contain no common element.

Consider now two disjoint sets W and V of expressions. A predicate H is said to *weakly* separate W from V if H represents some superset of W that is disjoint from V—in other words, if HX is provable for every X in W, but not provable for any X in V.

We say that H *strongly* separates W from V if H represents some superset of W, and H' represents some superset of V—in other words, if HX is provable for every X in W, and HX is refutable for every X in V.

W is said to be weakly (strongly) separable from V if some predicate weakly (strongly) separates W from V.

Problem 20. Which, if either, of the following statements is true?

(a) If W is weakly separable from V then V is weakly separable from W.

(b) If W is strongly separable from V then V is strongly separable from W.

Now for some Rosser type results.

Theorem R_1. If P is weakly separable from R, or if R is weakly separable from P, then there is an undecidable sentence.

More specifically, if H weakly separates P from R, or R from P, then HX is undecidable for some sentence X.

Problem 21. Prove Theorem R_1.

Problem 22. Theorem R_1 is really a strengthening of Theorems G and S. Why?

Theorem R_2. If H strongly separates P from R, or R from P, then HX is undecidable for some sentence X.

Problem 23. Prove Theorem R_2.

Theorem R [After Rosser, 1936]. If H strongly or even weakly, separates P* from R*, or R* from P* then HX is undecidable for some X.

Problem 24. Prove Theorem R.

What Rosser actually did was to strongly separate R* from P*.

Problem 25. Theorem R is stronger than Theorem R_1 (hence also stronger than R_2). Why?

To conclude this chapter, let me say that Gödel's proof essentially boils down to representing P*. In [Smullyan, 1961] I showed that an undecidable sentence can also be obtained by representing R*. Rosser achieved incompleteness by strongly separating R* from P*, and this could be done without assuming Gödels hypothesis of omega consistency.

Solutions to the Problems of Chapter XIV

1. Suppose H is a provability predicate. We first note that for any expression X:

(1) $H(H^{\#}\text{'}X)$ is provable iff $H(XX)$ is refutable.

Reason. $H(H^{\#}\text{'}X)$ is provable iff $H^{\#}\text{'}X$ is provable (since H is a provability predicate), which in turn is the case iff $H^{\#}X$ is refutable (by C_1), iff $H(XX)$ is refutable (by C_2).

Now comes Gödel's great trick: Since (1) is true for *every* expression X, it is true if X is the very predicate $H^{\#}\text{'}$. We then substitute $H^{\#}\text{'}$ for every occurrence of X in (1), and we get (2) $H(H^{\#}\text{'}H^{\#}\text{'})$ is provable iff $H(H^{\#}\text{'}H^{\#}\text{'})$ is refutable. Let S be the sentence $H(H^{\#}\text{'}H^{\#}\text{'})$. Thus S is provable iff S is refutable. This means that S is either both provable and refutable, or is neither provable nor refutable. Since we are assuming that the system is consistent, S cannot be both provable and refutable, hence S is neither provable or refutable. Thus S is undecidable.

This proves that the sentence $H(H^{\#}\text{'}H^{\#}\text{'})$ is undecidable. Thus HX is undecidable, for X equal to the expression $H^{\#}\text{'}H^{\#}\text{'}$.

For future reference, let us note once and for all that to show a given sentence to be undecidable, it suffices to show that the sentence is provable if and only if it is refutable.

Incidentally, the sentence $H(H\text{'}^{\#}H\text{'}^{\#})$ is also undecidable, as the reader can verify.

2. Again let H be a provability predicate.

(1) For any expression X, the sentence $H^{\#}\text{'}X$ is refutable iff XX is provable. *Reason*: $H^{\#}\text{'}X$ is refutable iff $H^{\#}X$ is provable, iff $H(XX)$ is provable, iff XX is provable (since H is a provability predicate). Again we take $H^{\#}\text{'}$ for X in (1), and we see that $H^{\#}\text{'}H^{\#}\text{'}$ is refutable iff $H^{\#}\text{'}H^{\#}\text{'}$ is provable, hence the diagonal sentence $H^{\#}\text{'}H^{\#}\text{'}$ is undecidable.

We remark that the diagonal sentence H'#H'# is also undecidable, as the reader can verify.

3. Suppose that H represents W and $\widetilde{W}$ is not representable. Then H' doesn't represent $\widetilde{W}$ (no predicate does). Hence there must be at least one expression X such that it is *not* the case that H'X is provable iff X is in $\widetilde{W}$ Thus H'X is provable iff X is *not* in $\widetilde{W}$, but to say that X is not in $\widetilde{W}$ is to say that X is in W. Thus H'X is provable iff X is in W. Thus HX is refutable iff X is in W. But also HX is provable iff X is in W (since H represents W). Therefore HX is refutable iff HX is provable, hence HX is undecidable.

4. Consider a refutability predicate K. For any expression X, the sentence K(K#X) is provable iff K#X is refutable (since K is a refutability predicate) iff K(XX) is refutable (by condition C_2)). Thus K(K#X) is provable iff K(XX) is refutable. We take K# for X, and so K(K#K#) is provable iff K(K#K#) is refutable. Thus K(K#K#) is undecidable.

5. Again consider a refutability predicate K. Then for any expression X, the sentence K#X is provable iff K(XX) is provable, iff XX is refutable. Thus K#K is provable iff XX is refutable. Again we take K# for X and so K#K# is provable iff K#K# is refutable, and so the diagonal sentence K#K# is undecidable.

6. Suppose K is a creative predicate. Since for *any* predicate H there is some X such that KX is provable iff HX is provable, we can take K' for H, and so there is an X such that KX is provable iff K'X is provable. Therefore KX is provable iff KX is refutable, and so KX is undecidable.

7. (a) Consider a provability predicate H and an arbitrary predicate K. By C_2, for any X, the sentence K#X is equivalent to K(XX). Taking K# for X, we have K#K# is equivalent to K(K#K#), thus K#K# is provable iff K(K#K#) is provable. But also K#K# is provable iff H(K#K#) is provable (since H is a provability predicate). Thus H(K#K#)

is provable iff $K^\#K^\#$ is provable, iff $K(K^\#K^\#)$ is provable. Hence $H(K^\#K^\#)$ is provable iff $K(K^\#K^\#)$ is provable, and thus HX is provable iff KX is provable, for X the sentence $K^\#K^\#$. Thus H is creative.

(b) Suppose H is a refutability predicate. We show that for any predicate K, the sentence HX is provable iff KX is provable, where X is the sentence $K^{\#'}K^{\#'}$.

Well, $H(K^{\#'}K^{\#'})$ is provable iff $K^{\#'}K^{\#'}$ is refutable (since H is a refutability predicate), which is true iff $K^\#(K^{\#'})$ is provable (by C_1), which is true iff $K(K^{\#'}K^{\#'})$ is provable. Thus HX is provable iff KX is provable, for X the expression $K^{\#'}K^{\#'}$.

[Incidentally another expression X such that HX is provable iff KX is provable is $X = K'^\#K'^\#$, as the reader can verify.]

8. For any expression X, the sentence $H^\#X$ is equivalent to H(XX). We take $H^\#$ for X, and so $H^\#H^\#$ is equivalent to $H(H^\#H^\#)$. Thus $H^\#H^\#$ is a fixed point of H.

9. (a) Suppose H is a provability predicate, K is an arbitrary predicate, and X is a fixed point of K. Then HX is provable iff X is provable (since H is a provability predicate), which is true iff KX is provable (since X is a fixed point of K).

(b) Suppose H is a refutability predicate, K is an arbitrary predicate, and X is a fixed point of K'. Then HX is provable iff X is refutable, which is true iff K'X is refutable (since X is a fixed point of K'), which is true iff KX is provable. Thus HX is provable iff KX is provable.

10. (a) Suppose H is a provability predicate and X is a fixed point of H'. Then HX is provable iff X is provable, which is true iff H'X is provable (since X is a fixed point of H'). Thus HX is provable iff H'X is provable, hence HX, and consequently also H'X, are undecidable. And since X is equivalent to H'X, X is also undecidable.

(b) Suppose K is a refutability predicate and X is a fixed point of K. Then KX is provable iff X is refutable, iff KX is refutable (since X is a fixed point of K). Thus KX is provable iff KX is refutable, and so KX is undecidable. Since X is equivalent to KX, then X is also undecidable.

11. (1) Suppose W is representable. Let H be a predicate that represents W.

(a) Let X be a diagonal fixed point of H, which must exist by Theorem F. Then X is provable iff HX is provable, iff X is in W (since H represents W).

(b) Let Y be a diagonal fixed point of H'. Then Y is refutable iff H'Y is refutable, iff HY is provable, iff Y is in W. Thus Y is refutable iff Y is in W.

11. (2) If there is a provability predicate then the set of provable sentences is representable, hence by (b) above, taking W to be the set of provable sentences, there is a diagonal sentence Y such that Y is refutable iff Y is provable, hence Y is undecidable.

If there is a refutability predicate, then the set of refutable sentences is representable, and so in (a) above, we take W to be the set of refutable sentences, and then there is a diagonal X which is provable iff X is refutable and so X is undecidable.

12. Both statements are true because:

(a) If the complement $\tilde{P}$ of P were representable, then by (a) of Theorem B, taking $\tilde{P}$ for W, there would be a sentence X which is provable iff X is in $\tilde{P}$, which would mean that X is provable iff X is not provable, which is impossible. Therefore $\tilde{P}$ is not representable.

(b) If $\tilde{R}$ were representable then by (b) of Theorem B, taking $\tilde{R}$ for W, there would be a sentence Y which is refutable iff Y is not refutable, which again is impossible.

13. To prove Theorem F_1, given a predicate H let K be the predicate $H^\#$. Then KK' is the sentence $H^\#K'$, which is

equivalent to H(K'K'). Thus KK' is equivalent to H(K'K'). Now by condition C_1, we know that K'K' is the negation of KK'—thus K'K' is the sentence $\overline{KK'}$. Thus KK' is equivalent to H($\overline{KK'}$), and so if we take X to be the skew diagonal sentence KK' (which is actually H*H*'), then X is equivalent to H($\overline{X}$).

As for the corollary, suppose W is representable. Let H be a predicate that represents W.

(a) Let X be a skew diagonal sentence that is equivalent to H($\overline{X}$). Then X is provable iff H($\overline{X}$) is provable, which is true iff $\overline{X}$ is in W (since H represents W). Thus X is provable iff $\overline{X}$ is in W.

(b) Let Y be a skew diagonal sentence that is equivalent to H'$\overline{Y}$. Then Y is refutable iff H'$\overline{Y}$ is refutable, which is true iff H$\overline{Y}$ is provable, which is true iff $\overline{Y}$ is in W. Thus Y is refutable iff $\overline{Y}$ is in W.

14. To prove Theorem G_2, suppose there is a provability predicate. Then the set of provable sentences is representable. In Corollary F_1 let W be the set of provable sentences. Then by (a) of Corollary F_1, there is a skew diagonal sentence X which is provable iff $\overline{X}$ is in W—thus X is provable iff $\overline{X}$ is provable, hence X is undecidable.

As for Theorem S_2, suppose the set of refutable sentences is representable. In (b) of Corollary F_1, take W to be the set of refutable sentences. Then there is a skew diagonal sentence Y such that Y is refutable iff $\overline{Y}$ is in W—thus Y is refutable iff $\overline{Y}$ is refutable, which is true iff Y is provable, and so Y is undecidable.

15. It is (b) that is true. If the skew diagonal sentence KK' is undecidable, so is the diagonal sentence K'K'.

16. Suppose W* is representable. Let H be a predicate that represents W*.

(a) For any expression X, the sentence HX is provable iff XX is in W (because HX is provable iff X is in W*, iff

XX is in W). Take H' for X, and so HH' is provable iff H'H' is in W. However H'H' is the negation of HH', and so X is provable iff its negation $\overline{X}$ is in W, where X is the skew diagonal sentence HH'.

(b) This is a bit trickier: Since H represents W*, so does H", as the reader can easily verify. Thus for any expression X, the sentence H"X is provable iff XX is in W. We now take H" for X, and so H"H" is provable iff H"H" is in W. Now H"H" is provable iff H'H" is refutable and so H'H" is refutable iff H"H" is in W. To reduce clutter, let K be the predicate H'. Thus KK' is refutable iff K'K' is in W, but K'K' is the negation of KK'. Thus the skew diagonal sentence KK' is refutable iff its negation $\overline{KK'}$ is in W.

17. Suppose P* is representable. In (a) of Theorem W we take P for W, and so there is a skew diagonal sentence X such that X is provable iff $\overline{X}$ is in P, which means that X is provable iff $\overline{X}$ is provable, and hence X is undecidable. Since some skew diagonal sentence is undecidable, so is some diagonal sentence, by Problem 15.

Next, suppose that R* is representable. Then by (b) of Theorem W, taking R for W, there is a skew diagonal sentence Y such that Y is refutable iff $\overline{Y}$ is refutable, hence Y is refutable iff Y is provable, and so Y is undecidable. Again, there must also be an undecidable diagonal sentence.

18. It is (a) that is true. If H represents W then $H^\#$ represents W*, since for all X, the sentence $H^\#X$ is provable iff H(XX) is provable, which is true iff XX is in W (since H represents W), which is true iff X is in W*.

19. Since the representability of a set W implies the representability of W*, as we have just seen, the hypothesis of either Theorem G_1 or G_2 implies the hypothesis of Theorem G_0^*, and the hypothesis of either Theorem S_1 or Theorem S_2 implies the hypothesis of Theorem G_0^*.

Likewise the hypothesis of Corollary F_1 implies the hypothesis of Theorem W.

20. It is (b) that is true. If H strongly separates W from V then H' strongly separates V from W, for suppose H strongly separates W from V. Then for any X:

(a) If X is in V, then HX is refutable, hence H'X is provable.

(b) If X is in W, then HX is provable, hence H'X is refutable.

21. (a) Suppose H weakly separates P from R. Thus for every expression X,

(1) If X is provable (X is in P) then HX is provable

(2) If HX is provable then X is not refutable (X is not in R).

Now let X be a fixed point of H'. Then if X is provable, so is H'X. But by (1), if X is provable so also is HX, and the system is then inconsistent. Since we are assuming the system to be consistent, then X is not provable.

Suppose X is refutable. Then H'X, being equivalent to X, is also refutable, and hence HX is provable, and then by (2), X is not refutable, and we have a contradiction! Thus X is not refutable. Hence X is undecidable. Since X is equivalent to H'X, then H'X is also undecidable, hence so also is HX.

This proves that if H separates P from R then HX is undecidable for some X—in fact for any X that is a fixed point of H'.

(b) Suppose H weakly separates R from P. Thus for any X,

(1) If X is refutable, then H(X) is provable.

(2) If X is provable, then H(X) is not refutable.

Now let X be a fixed point of H.

If X is refutable, then H(X) is provable (by (1)), but also refutable (since X is a fixed point of H) and the system is then inconsistent.

Hence, assuming consistency, X is not refutable.

If X is provable, then by (2), H(X) is not refutable, but also H(X) is refutable (since X is a fixed point of H) and we then have a contradiction. Thus X is not provable, and therefore X is undecidable, and so is the equivalent sentence H(X).

22. Because if H represents P then it certainly weakly separates P from R, since it represents a superset of P (namely P itself) that is disjoint from R. Thus the hypothesis of Theorem G implies the hypothesis of Theorem R_1. Similarly the hypothesis of Theorem S implies the hypothesis of Theorem R_1.

We might also note that the proof we gave of Theorem R_1 revealed further information, namely that not only HX, but X itself is also undecidable, where X is any fixed point of H' if H represents P, or of H itself, if H represents R. By taking X to be a skew diagonal fixed point, we get strengthenings of Theorems G_2 and S_2 as well (and hence also of G_1 and S_1).

23. We could easily prove this from scratch, but the fact is that if H strongly separates a set W from a set V then it also weakly separates W from V. Reason: Suppose H strongly separates W from V. Then for all X,

1. If X is in W then HX is provable.

2. If X is in V then HX is refutable, hence also HX is not provable (since we are assuming the system to be consistent).

Thus Theorem R_2 is a consequence of Theorem R_1.

24. (a) Suppose H weakly separates R* from P*. Then for any expression X,

(1) If XX is refutable then HX is provable (because if XX is refutable, XX is in R, hence X is in R*, hence HX is provable (since H weakly separates R* from P*).

(2) If XX is provable then HX is not provable (because if XX is provable, then XX is in P, hence X is in P*, hence HX is not provable, since H weakly separates R* from P*.

In (1) and (2) we take X to be H itself, and so

(1)' If HH is refutable then HH is provable.

(2)' If HH is provable then HH is not provable.

By (1)', if HH is refutable then the system is inconsistent, and since we are assuming the system to be consistent, HH is not refutable.

By (2)' if HH is provable, we get a logical contradiction, hence HH is not provable. Thus HH is undecidable.

(b) Suppose H weakly separates P* from R*. Then for any expression X:

(1) If XX is provable then HX is provable.

(2) If XX is refutable then HX is not provable.

In (1) and (2) we now take H' for X, and so:

(1)' If H'H' is provable then HH' is provable.

(2)' If H'H' is refutable then HH' is not provable.

Now, if HH' is refutable then H'H' is provable, hence by (1)', HH' is also provable and the system is inconsistent. Therefore HH' is not refutable.

If HH' is provable then H'H' is refutable, hence by (2)', HH' is not provable, and we have a logical contradiction. Therefore HH' is not provable. Thus HH' is undecidable.

25. Because if for any two sets W and V of expressions, if W is weakly separable from V, then W* is weakly separable from V*—more specifically, if H weakly separates W from V, then H# weakly separates W* from V*, because suppose H weakly separates W from V. Then for any X:

(1) If X is in W* then XX is in W, hence H(XX) is provable, hence H#X is provable.

(2) If X is in V* then XX is in V, hence H(XX) is not provable, hence H#X is not provable.

This shows that if H weakly separates W from V, then $H^\#$ weakly separates W* from V*. [Incidently, it is also true that if H strongly separates W from V, then $H^\#$ strongly separates W* from V*, as the reader can easily verify].

Now, to see that Theorem R is stronger than Theorem R_1, suppose the hypothesis of Theorem R_1 holds—there is a predicate H that either weakly separates P from R or R from P. Then $H^\#$ either separates P* from R*, or R* from P*, hence by Theorem R applied to the predicate $H^\#$, there is some X such that $H^\#X$ is undecidable. Since $H^\#X$ is equivalent to H(XX), then H(XX) is undecidable, hence there is an expression Y—namely XX—such that HY is undecidable, which proves Theorem R_1.

CHAPTER XV
PROVABILITY IN STAGES

Part I. Re Gödel's Proof

§1. *Omega Consistency.* We have already remarked that Gödel proved the existence of an undecidable sentence [Gödel, 1931] only by assuming the system had a stronger property than consistency—a property known as omega consistency. Before discussing this notion precisely, I want to make some purely informal remarks to serve as a heuristic background.

Imagine that we are all immortal, but there is a sleeping sickness which if caught would put one to sleep forever. However there is an antidote which administered would wake the sleeper up, but only for a limited time. The problem now is this: Suppose that your loved one contracts the sleeping sickness today.

If you give her the antidote today, she will wake up for 2 days and then go back to sleep forever. If you give the antidote on the next day, she will wake up for 4 days. . . If you give her the antidote in n days from now, she will wake up for 2^n days. Now, you wish her to be with you for as many days as possible, but on any one day, if instead of giving her the antidote, you wait just one more day, you will have her for twice as long! Thus, on any one day it is

irrational to give it on that day, yet it is certainly irrational never to give her the antidote at all!

This disturbing situation might be considered an
example of an omega consistency, which in mathematics
roughly means the following: Consider a mathematical
system about in which there is a certain property such
that there is a proof that 1 doesn't have the property,
another proof that 2 doesn't have the property, and for
each number n there is a proof that n doesn't have the
property, and at the same time there is a proof that there
exists a number that does have the property!

Such a system would be called omega-inconsistent.
Despite the oddness of such a system, one cannot derive a
formal inconsistency from it. There are indeed consistent
systems that are omega inconsistent. The point is that a
proof consists of only a *finite* sequence of sentences and
so given an *infinite* sequence of sentences, even though it
may be impossible that all of them are true, one cannot
necessarily demonstrate this with any *finite* number of
the sentences.

The situation can be nicely analogized as follows: Imagine that we are all immortal and that there are infinitely
many banks in the universe—Bank 1, Bank 2, ... Bank n.
. . . You get a check saying PAYABLE AT SOME BANK.
You take it successively to Bank 1, Bank 2, ... Bank n ...,
and at no time have any of the banks honored it. Even
after you have tried billions of banks, you cannot be sure
that the next bank you try may not honor it and so, at
no time can you prove that the check is invalid. Now, if
there had been only finitely many banks in the universe,
then after having tried them all and failed with each one,
you would have proved that the check is invalid. But with
infinitely many banks, you can never prove the check

invalid, even though it is. This is an example of an omega inconsistentcy.

A humorous illustration of an omega inconsistentcy was given by the mathematician Paul Halmos: He defined an *omega inconsistent mother* as a mother who says to her child, "You may not do *this*, you may not do *that*, you may not do —, ..." The child asks, "Isn't there something I can do?" The mother replies, "Yes, there is *something* you can do, but its not this, nor that, nor ...

Provability in Stages. In the systems under consideration, there is a well defined relation between sentences and numbers, which is read "X is provable at stage n."

[I might remark that in application to actual systems, there are certain finite sequences of expressions called *proofs*, and a sentence X is provable if and only if it is the last line of some proof, and such a proof is called a proof of X. Now, Gödel assigned numbers, not only to expressions but also to proofs, and if I too assigned Gödel numbers to proofs, I could then say that X is provable at stage n to mean that n is the Gödel number of a proof of X. But for now, I want the notion "X is provable at stage n" to be quite general.)

In the systems under consideration, for each number n there is a predicate S_n, where for each expression X, the sentence S_nX is interpreted to mean that X is provable at stage n. These predicates satisfying the following two conditions (for all sentences X).

G_1. If X is provable at state n then the sentence S_nX is provable.

G_2. If X is not provable at stage n then the sentence S_nX is refutable.

[The idea behind conditions G_1 and G_2 is that the system, so to speak, has at each stage perfect memory for what it has so far proved and not proved.]

In addition there is a predicate Q, where for any expression X, the sentence QX is interpreted to mean that for all numbers n, X is not provable a stage n—in other words, X is not provable. Thus QX can be thought of as a *universal* statement, since it is a statement about all numbers n—namely that S_nX doesn't hold. [In the notation of first-order logic, QX would be written: $\forall n \sim S_nX$—read "For all n, not S_nX." The symbol "$\forall$" is called the *universal* quantifier.]

Under this interpretation Q'X would mean that it is *not* the case that S_nX is false for all n, hence S_nX is true for at least one n—in other words there *exists* an n such that X is provable at stage n. Accordingly S'X is to be thought of as an *existential* statement (and in the notation of first order logic could be written: $\exists n \, S_nX$, read "there exists an n such that S_nX holds" The symbol "$\exists$" is called the *existential quantifier*, and is read "there exists") Under this interpretation, if there is at least one n such that S_nX is true, then Q'X is true, and this is formally reflected by the fact that in the systems under consideration, the following condition holds:

Condition E [E for *existential*]. If for some n, the sentence S_nX is provable, then Q'X is provable.

We are *not* given that the converse of condition E holds. If Q'X is true (under the intended interpretation), then of course S_nX must be true for at least one number n, but that does not necessarily mean that if Q'X is provable, then S_nX must be provable for some n! However, if the existential statement Q'X is provable and if S_nX is *refutable* for every n, then the system would be called *omega inconsistent*. Thus if the system is omega consistent, then if Q'X is provable, there must be at least one number n such that S_nX is *not* refutable.

To avoid possible confusion, what we have previously called *consistent* (no sentence and its negation are both provable) we should now call *simply consistent* (in contradistinction to omega consistency). This practice is generally followed in the literature.

We continue to assume conditions C_1 and C_2 of the last chapter. From these and the three conditions G_1, G_2 and E, we have the following key result:

Theorem G [*Gödel's Incompleteness Theorem*, Gödel, 1931]. If the system is omega consistent (as well as simply consistent) then there is an undecidable sentence.

More specifically, there is a sentence G such that:

(a) G is not provable (assuming simple consistency);

(b) If the system is omega consistent, then G is not refutable.

Problem 1. Prove Theorem G.

Problem 2. Assuming the system to be omega consistent, which, if either, of the following statements are true?

(a) Q is a provability predicate.

(b) Q' is a provability predicate.

Problem 3. For the sentence G constructed in the solution of the last problem, although neither QG of Q'G is provable in the system, one of those two sentences must be true under the intended interpretation). Which one is the true one?

Omega Incompleteness. A system is called *incomplete* (sometimes *simply incomplete*) if some sentence is undecidable. Gödel's famous theorem about the existence of undecidable sentences is known as *Gödel's incompleteness theorem.*

The systems subject to Gödel's arguments have an even more surprising property known as *omega-incompleteness.* Suppose we have a certain property of natural numbers and that there is a proof that 0 has the property, another proof that 1 has the property, and for every natural number n, there is a proof in the system that n has the property, yet there is no proof in the system of the single statement "All numbers have the property." Such a system is called *omega-incomplete*. Well, the systems under present consideration have that strange character!

In the solution we gave proving Theorem G, the sentence QG, as well as G itself, is undecidable, hence not provable (though it is true under the intended interpretation). Since G is not provable, then for every number n, G is not provable at stage n, hence by condition G_2, the sentence S_nG is refutable. Thus for every number n, the sentence $S_n'G$ is provable. Yet the universe sentence QG, which asserts that for all n, G is not provable at stage n, —this sentence is not provable!

[In the notation of first-order logic, the sentence QG could be written $\forall_n S_n'G$. Thus all the infinitely many sentences $S_0'G$, $S_1'G$, $S_2'G$, ..., $S_n'G$, ... are provable, yet the universal sentence $\forall n S_n'G$ is not provable!] This is a case of omega incompleteness.

Virtually all mathematicians have heard of Gödel's incompleteness theorem, though few are familiar with the proof. But most mathematicians have not even *heard* of the fact that the systems under consideration are omega incomplete, and those that I have informed of this have found it even more surprising than the fact that these systems have undecidable sentences. Yet the omega incompleteness easily follows from Gödel's proof that there are undecidable sentences.

Part II. Re Rosser's Proof

In the last chapter we proved that if some predicate strongly separates R from P and the system is simply consistent, then the system contains an undecidable sentence. Rosser did manage to construct a predicate that did that. I now wish to give you a sketch of how Rosser accomplished this. It can be only a sketch, since a more adequate account can be given only within the framework of systems using first-order logic. A more complete description will be given later on, but what we now do in this chapter will provide a good basis for what we will do later.

Provability By Stages. We shall say that a sentence is provable *by* stage n to mean that it is provable at stage n or at some earlier stage—in other words that there is some number m less than or equal to n such that the sentence is provable at stage m.

In addition to the predicates S_0, S_1, S_2, ..., S_n, we now also have predicates B_0, B_1, B_2, ..., B_n, ... where for each number n, the sentence $B_n X$ is interpreted to mean that X is a sentence that is provable *by* stage n. In addition to the conditions G_1, G_2 already considered, the systems under consideration satisfy the following two conditions:

G_3. If for some m less than or equal to n, the sentence $S_m X$ is provable, then $B_n X$ is provable.

G_4. If for every m less than or equal to n, the sentence $S_m X$ is refutable, then $B_m X$ is refutable.

We no longer need Gödel's predicate Q, nor condition E, nor need we now consider the question of omega consistency. Instead of Gödel's predicate Q, in Rosser's proof we have a predicate that I will denote by "J" (after J. Barkley Rosser) satisfying conditions J_1, J_2, soon to be given. For any expression X, the sentence JX is interpreted to mean that X is a sentence such that for all numbers n, either X is not provable at stage n, or X is refutable *by* stage n.

To motivate the conditions J_1, J_2, shortly to be given, we had best first consider the following problem:

Problem 4. Consider two sentences X and Y and the following proposition: "For all numbers n, either X is not provable at stage n, or Y is provable by stage n."

(a) Suppose there is a number m such that X is provable at stage m and Y is not provable by stage m. Is the above proposition true or false?

(b) Suppose on the other hand that it is Y that is provable at stage m and X is not provable by stage m. Is the above proposition then true or false?

Note: The answer to (a) is quite trivial. As for (b), the answer is not at all obvious, and involves a very cute trick!

We see from the above problem, taking $\overline{X}$ for Y, that for any number m:

(a) If X is provable at stage m and $\overline{X}$ is not provable by stage m, then JX is false (under the intended interpretation)—i.e., it is not the case that for all numbers n, either X is not provable at stage n or $\overline{X}$ is provable by stage n.

(b) If $\overline{X}$ is provable at stage n and X is not provable by stage n, then JX is true.

Thus, under the intended interpretation of all the predicates S_n, B_n, as well as J, the following holds:

(1) If $S_m X$ is true and $B_m \overline{X}$ is false, then JX is false.

(2) If $S_m \overline{X}$ is true and $B_m X$ is false then JX is true.

The above two facts are formally reflected by the fact that the predicate J satisfies the following two conditions: (for every sentence X and any number m).

J_1: If $S_m X$ is provable and $B_m \overline{X}$ is refutable, then JX is refutable.

J_2: If $S_m \overline{X}$ is provable and $B_m X$ is refutable, then JX is provable.

[Note that conditions J_1 and J_2 are respectively facts (1) and (2) above, replacing "true" by "provable," and "false" by "refutable."]

The stage is now set, and we have:

Theorem R [After Rosser]. In any simple consistent system satisfying conditions C_1, C_2, G_1, G_2, G_3, G_4, J_1, J_2, there is a sentence X such that X and JX are undecidable.

Problem 5. Prove theorem R.

Discussion. Gödel's undecidable sentence can be thought of saying, "I am not provable" Rosser's curious sentence can be thought of as saying:

"At any stage n, either I am not provable at stage n, or I am refutable at that or some earlier stage."

I must tell you of a very interesting incident. Professor Rosser and his lovely wife Annette visited my late wife Blanche and me about ten years ago in our house in the beautiful upper Catskills. They stayed for several days. I remember dear old Rosser coming downstairs in his dressing gown and leisurely lounging around until dinner. At one point I asked him how he ever discovered his curious sentence that enabled him to prove incompleteness without having to assume Gödel's hypothesis of omega consistency. He replies, "I was simply experimenting with several variations of Gödel's sentence, and when I came to this one, I suddenly realized what I could do with it!"

I found that most interesting! It reminded me of a rumor I heard that Gödel originally meant to prove the systems under consideration not incomplete but inconsistent! He thought that he could create the famous liar paradox ("this sentence is false") within the system. He evidently thought that he could define truth in the system, not yet realizing that truth was not the same as provability, which is what in fact he did capture.

Solutions to the Problems of Chapter XV

1. We take G to be a fixed point of Q.

(a) Suppose G is provable. Then it is provable at some stage n. Then by condition G_1, the sentence S_nG is provable. Then by condition E, the sentence Q'G is provable, and thus QG is refutable. And so, if G is provable, then QG is refutable. But since G is a fixed point of Q, then if G is provable, so is QG. Therefore if G is provable, then QG is both refutable and provable, which means the system is simply inconsistent. Since we are assuming the system to be consistent, then G is not provable.

(b) Now suppose the system is omega consistent (as well as simply consistent). Since G is not provable (by (a)), it is not provable at any stage. Hence by condition G_2, the sentence S_nX is refutable for every number n. Then, since we are assuming omega consistency, the sentence Q'G is not provable. Thus QG is not refutable, and since G is equivalent to QG, then G is not refutable. Hence G (as well as QG) is undecidable.

2. Assuming the system to be omega consistent, it is Q' that is a provability predicate, as we can see as follows:

(1) Even without assuming omega consistency, if X is provable, so is Q'X, because X is then provable at some stage n, hence S_nX is provable, and therefore Q'X is provable.

(2) For the converse (if Q'X is provable, so is X) we need to assume omega consistency:

Suppose Q'X is provable. If X is not provable, then S_nX is refutable for every n, which would mean the system is omega inconsistent, contrary to our assumption that the system is omega consistent. Thus X is provable.

3. Obviously QG, which asserts that G is not provable, since in fact G is not provable.

4. (a) This is indeed trivial: Suppose X is provable at stage m and Y is not provable by stage m. Then there is at least one number n, namely m itself such that it is *not* the case that either X is not provable at stage n or Y is provable by stage n, since for such an n, X *is* provable at stage n and Y is *not* provable by stage n. Thus it is false that for *all* n, either X is not provable at stage n or Y is provable by stage n.

(b) This case is more interesting: Suppose Y is provable at stage m and X is not provable by stage m. To show that the proposition in question is true, the trick is to divide the proof into two cases—the case that $n \leq m$ (n is less than or equal to m) and the case that $m \leq n$ (m is less than or equal to n).

Case 1. $n \leq m$. In this case, since X is not provable by stage m and $n \leq m$, then X is not provable at stage n.

Case 2. $m \leq n$. In this case, since Y is provable at stage m and $m \leq n$, then Y is provable *by* stage n.

Thus either X is not provable at stage n (Case 1) or Y is provable by stage n (Case 2).

5. We will show that J strongly separates R (the set of refutable sentences) from P (the set of provable sentences), and the result will then follow from Theorem R_2 of the last chapter.

Thus we are to show that for any sentence X, if X is refutable then JX is provable, and if X is provable, then JX is refutable.

(1) Suppose X is refutable. Then $\overline{X}$ is provable. Then $\overline{X}$ is provable at stage m for some m, hence by condition G_1, the sentence $S_m\overline{X}$ is provable. Since X is refutable, and we are assuming the system to be simply consistent, then X is not provable, hence not provable at any stage, hence not provable by stage m, and therefore by conditions G_2 and B_0, the sentence $B_m X$ is refutable. Thus $S_m\overline{X}$ is provable

and $B_m X$ is refutable, and so by condition C_1 of Chapter XIV and condition J_2, the sentence JX is provable. This proves that if X is refutable, then JX is provable.

(2) Suppose X is provable. Then it is provable at some stage m, hence $S_m X$ is provable. Since X is provable, then by the assumption of simple consistency, $\overline{X}$ is not provable, hence not provable by stage m, and so by condition G_2, the sentence $B_m \overline{X}$ is refutable. Thus $S_m X$ is provable and $B_m \overline{X}$ is refutable, and so by condition J_1, the sentence JX is refutable. Thus if X is provable, then JX is refutable.

CHAPTER XVI

FORMAL SYSTEMS AND RECURSION

In the last three chapters we considered various conditions for a mathematical system to have an undecidable sentence. An equally, if not more, important question about these systems is whether there is a purely mechanical way of determining which sentences are and are not provable in the systems. Still more important, in the present writer's opinion (which is not universally shared), is whether there is a purely mechanical method of determining which of the sentences of the system are *true*. Since in these systems, truth and provability don't coincide, which class of sentences is more important—the class of provable sentences or the class of true sentences? Surely the overwhelming bulk of mathematicians are more interested in which sentences are true. Consider, for example, Goldbach's famous conjecture that every even number greater than 2 is the sum of two prime numbers. To this day, no one knows whether it is true or not. Surely any mathematician working in number theory is less interested in whether Goldbach's conjecture is provable in some given mathematical system, than whether it is true—i.e. whether every even number greater than 2 really is the sum of two primes!

But now, to answer questions about whether or not there is a mechanical procedure to solve a given problem, we must first define exactly what is meant by a *mechanical* procedure. Informally, a mechanical method is one that can be carried out without any creative ingenuity—one that can be carried out by a machine. But this informal characterization is not good enough for strictly mathematical purposes. So how can we more accurately define what is meant by a *mechanical* procedure? Well, many proposed definitions have been given by various authors. In these definitions the mechanical procedures being defined have been called the following: recursive, Turing computable, Abas computable, Internal Register Machine computable, Markov computable, computable definable, definable in combinatory logic, definable in a Post canonical system, definable in an elementary formal system, as well as many, many other names. The interesting thing is that these various characterizations, which appear so different on the surface, have all turned out to be equivalent! That is, a set is recursive if and only if it is Turing computable, if and only if it is Abas computable, if and only if . . . if and only if it is definable in an elementary formal system. This constitutes strong heuristic evidence that these various characterizations have really captured what is meant by a *mechanical* process. The approach that I will now take is that of *elementary formal systems*, which I introduced in [Smullyan, 1961].

I will first define a certain type of program that I will call an *elementary* program. Before giving a general definition I will give several examples.

Let us first consider the set of all expressions using just two symbols a and b—expressions like abbaaaba, or babbbbaab—any string of a's and b's. Let x and y stand for any such expressions. By xy we mean the expression x immediately followed by y. For example if x is the expression aab and y is the expression bab, then xy is the expression aabbab. Or if x is the expression aab then by xbb is meant aabbb, and xa is the expression aaba.

Now suppose we want to generate the set A of all strings in a and b that contain no two consecutive occurrences of a or of b—call such strings *alternating*. The following facts clearly hold:

1. a is alternating
2. b is alternating
3. ab is alternating
4. ba is alternating
5. If xa is alternating, so is xab (x is any string)
6. If xb is alternating, so is xba

Also, no string is alternating unless its being so is a consequence of the above six conditions. We thus give the following instructions to generate the set A:

1. Put a in A
2. Put b in A
3. Put ab in A
4. Put ba in A
5. For any x, if xa is in A, then put xab in A
6. For any x if xb is in A, then put xba in A.

Now, programs are to be written in a symbolic language. A phrase such as "put x in A" is simply written "Ax." Thus 1, 2, 3, 4, are now written:

(1) Aa
(2) Ab
(3) Aab
(4) Aba

The if-then relation is symbolized by "→." Thus (5) and (6) are now written:

(5) Axb → Axba
(6) Axa → Axab

The letter "x" is used as a variable standing for any string of a's and b's. Given an expression involving the variable "x," by an *instance* of the expression is meant the result of substituting some string of a's and b's for every occurrence of "x" in the expression (the same string for every occurrence of x). For example, in (5) (Axb → Axba) an instance of it is the result of substituting aaba for x, which is Aaabab → Aaababa. Or again, an instance of (6) is Aaa → Aaab (by substitution a for x).

Now, here is how the set A gets generated. By (1), the single letter a gets put in A. By (2), (3), (4), we see that b, ab and ba get put in A. Now, in (5), if we replace x by a, we have the instance Aab → Aaba, which reads "if ab gets put in A, so does aba," but we already know that ab has been put in A, and therefore aba gets put in A. Once we have aba in A, we then put in abab (by (6), taking x to be aba). Then we successively get ababa, ababab, etc. (by repeated use of (5) and (6) we then get all alternating strings that begin with a. Also, since b gets in A, so does ba, bab, baba, ..., and so we thus also get all alternating strings beginning with b. And so we eventually get all alternating strings put in A.

In an instruction of the form x → y, the x is called the *antecedent* and y the *consequent*. For example, in (5) the antecedent is Axb and the consequent is Axba.

The above is an example of an elementary program (also called an elementary formal system in [Smullyan, 1961]). More specifically, this program would be called an elementary program over the alphabet {a, b}—the alphabet consisting of just the two symbols a and b. The symbol "A" is an example of what is called a *predicate* and is said to *represent* the set of all alternating strings.

Elementary programs also provide means of representing relations between strings. For example let K be the

3-symbol alphabet {a, b, c}. By the reverse of a string of these three symbols is meant the string in which the symbols are written in reverse order. For example, the reverse of cabbab is babbac. This relation of reversal is completely determined by the following conditions:

(1) The symbol a alone is the reverse of itself
(2) b is the reverse of itself
(3) c is the reverse of itself
(4) If x is the reverse of y then ax is the reverse of ya
(5) If x is the reverse of y then bx is the reverse of yb
(6) If x is the reverse of y then cx is the reverse of yc

We use the symbol "R" as a name of the reverse relation, and we wish to generate statements of the form Rx, y, whenever x is the reverse of y, and never when x is not the reverse of y. The following instructions accomplish this:

(1) Ra, a
(2) Rb, b
(3) Rc, c
(4) $Rx, y \rightarrow Rxa, ay$
(5) $Rx, y \rightarrow Rxb, by$
(6) $Rx, y \rightarrow Rxc, cy$

This program can be shortened a bit by observing the following facts:

(1) a is a single symbol
(2) b is a single symbol
(3) c is a single symbol
(4) If x is a single symbol, then x is its own reverse
(5) If x is the reverse of y and if z is a single symbol, then xz is the reverse of zy.

Thus we abbreviate "x is a single symbol" by "Sx", and have the following program:

(1) Sa
(2) Sb
(3) Sc

(4) $Sx \rightarrow Rx, x$

(5) $Rx, y \rightarrow Sz \rightarrow Rxz, zy$

We are here using the symbol "$\rightarrow$" as implication with *association to the right*—that is, for any statements X, Y and Z, the statement $X \rightarrow Y \rightarrow Z$ is read "if X is true, then Y implies Z," or "If X is true, then if Y is also true, so is Z," and *not* read "If X implies Y, then Z is true" (which would be association to the left). Similarly, for any four statements X, Y, Z and W, the statement $X \rightarrow Y \rightarrow Z \rightarrow W$, is read "if X is true, then if Y is true, then if Z is true, so is W," which could also be read "If X, Y and Z are all true, so is W," and could alternatively be symbolized: $(X \& Y \& Z) \rightarrow W$, where "&" is the symbol for "and." For technical reasons, it is best not to use the extra logical connection "&," but only the logical connective "$\rightarrow$" for "implies."

Here is another program for the reverse relation, which is still shorter and also works:

(1) Ra, a

(2) Rb, b

(3) Rc, c

(4) $Rx, y \rightarrow Rz, w \rightarrow Ryz, wx$

Before proceeding further, I should now give a *precise* definition of an *elementary formal system* [Smullyan, 1961].

By an *alphabet* K we shall mean a finite set of elements called the *symbols, signs,* or *letters* of K. Any finite sequence of symbols of K is called a *string*, or an *expression* or a *word* in K, or more briefly a *K-string*. For any K-strings X and Y, by XY is meant the sequence X followed by the sequence Y—for example if X is the string am, and Y is the string hjkd, then XY is the string amhjkd. The string XY is called the *concatenation* of X and Y.

By an *elementary formal system* (E) over K we mean a collection of the following items:

(1) The alphabet K.

(2) Another alphabet of symbols called *variables*. We will usually use the letters x, y, and z, with or without subscripts as our variables.

(3) Still another alphabet of signs called *predicates*, each of which is assigned a positive integer called the *degree* of the predicate. We usually use capital letters for predicates.

(4) Two more symbols called the *punctuation* sign (usually a comma) and the *implication sign* (usually "→").

(5) A finite sequence of strings which are *formulas*, according to the definition given below.

First some preliminary definitions: By a *term* we mean any string composed of symbols of K and variables. For example, if a, b, c are symbols of K, and x, y are variables, then aycxxbx is a term. So is xyyxx, and so is bcabbc. A term without variables will be called a *constant* term. By an *atomic formula* we mean an expression Pt, where P is a predicate of degree 1 followed by a term t, or an expression Rt_1, t_2, where R is a predicate of degree 2, and t_1 and t_2 are terms, or more generally for any positive integer n, a predicate of degree n followed by n terms separated by commas. If F and G are formulas, then F → G is a formula.

By a *sentence* we shall mean any formula without variables.

By an *instance* of a formula, we shall mean the result of substituting strings in K for occurrences of all the variables of the formula, with the understanding that if a variable has more than one occurrence in the formula, the same string in K must be substituted for each of the occurrences of the variable. For example, consider a formula Paxbycx, where a, b, c are symbols of K and x and y are variables. Suppose we substitute ab for x, and ca for y. We then get the instance Paabbcacab. If a formula has no variables—i.e., if it is a sentence—its one and only instance is itself.

An elementary formal system—denoted (E)—also has of a finite set of distinguished formulas called the initial formulas, or *axiom schemes* of the system. The set of all instances of all the axiom schemes are called the *axioms*, or *initial sentences* of the system.

We now define a sentence to be *provable* in the system if its being so is a consequence of the following two conditions:

(1) Every axiom of the system is provable in the system.

(2) For every *atomic* sentence X and any sentence Y, if X is provable and X → Y is provable, then Y is provable.

More precisely, by a *proof* in the system (E) is meant a finite sequence of sentences of (E) (usually displayed vertically, rather than horizontally), called the *lines* of the proof, such that for each line Y of the proof, either Y is an axiom of (E), or there is an *atomic* sentence X such that both X and X → Y are earlier lines of the proof.

A sentence is then called *provable* in (E) if it is the last line of some proof, and such a proof is called a proof of X.

Representability. For any elementary formal system (E) over an alphabet K, a predicate P of degree 1 is said to *represent* the set of all *constant* terms t such that Pt is provable in the system. A set S of constant terms of (E) is said to be representable in (E) if some predicate of (E) represents it.

A relation R(x, y) will be said to be represented by a predicate P of degree 2 if for all constant terms t_1 and t_2, the relation $R(t_1, t_2)$ holds if and only if the sentence Pt_1, t_2 is provable in the system. More generally, a relation $R(x_1, ..., x_n)$ of n arguments will be said to be represented by a predicate P of degree n if for all constant terms t_1, ..., t_n, the relation $R(t_1, ..., t_n)$ holds if and only if the sentence $Pt_1, ..., t_n$ is provable in the system.

A set or relation is said to be formally representable over K or *K-representable* if it is represented in some elementary formal system over K. Finally, a set or relation is said to be *formally representable* if it is formally representable in some elementary formal system over some alphabet K.

Proposition 1. Suppose that S_1 and S_2 are both sets of K-strings, and S_1 and S_2 are both K-representable, say respectively by elementary systems (E_1) and (E_2) [which could be the same]. Then there is an elementary formal system (E) over K in which S_1 and S_1 are both representable.

Problem 1. Prove Proposition 1 (this should be pretty obvious!)

For any two sets S_1 and S_2, by their *union* (symbolized by $S_1 \cup S_2$) is meant the set of all things that are either in S_1 or in S_2 (or both). By the *intersection* of S_1 and S_2 (symbolized by $S_1 \cap S_2$) is meant the set of all things that are in both S_1 and S_2.

Problem 2. Suppose S_1 and S_2 are both K-representable. Prove that $S_1 \cup S_2$ and $S_1 \cap S_2$ are both K-representable.

For any set S of K-strings, by the *complement* of S (with respect to K) is meant the set of all K-strings that are *not* in S. A set of K-strings will be said to be *solvable* over K, or *K-solvable* if both it and its complement are K-representable. A set will be called *solvable* if it is K-solvable over some alphabet K.

Discussion. The word *solvable* is well chosen: Given an elementary formal system (E) one can easily program a computer to generate the set of all sentences provable in (E). Now suppose a certain set S of K-strings is solvable over K. Thus S and its complement are each

K-representable. Then by Proposition 2, there is an elementary formal system (E) in which S is represented by some predicate—say "P"—and its complement is represented by some predicate—say "Q." Suppose we want to know whether or not a certain K-string X is or is not in S. There is then a perfectly effective way of finding out: Start a computer going to generate the provable sentences. If X is in S, then sooner or later the computer will print out PX, and if X is not in S, then sooner or later the computer will print out QX, and thus, sooner or later, we will know whether or not X is in S (assuming we live long enough.)

Now, suppose a set S is representable but not solvable—say it is represented in some elementary formal system by some predicate P. We would like to know of a given string whether or not it is a member of S. The best we can then do is to start a computer going to print out the provable sentences of the system. If X does belong to the set S, then sooner or later the computer will print out PX and we will know that X does belong to S, but if X does not belong to S, then the computer may run on forever, and at no stage can we know whether or not PX will be printed in the future. In short, if S is representable but not solvable, if X belongs to S, we will eventually know it, but if X doesn't belong to S, then we will never know it (unless by some creative ingenuity, we discover a way of finding out). Such a set S might aptly be called *semi-solvable*.

Do there exist sets which are semi-solvable but not solvable? That is a fundamental question in recursion theory and will be answered later in this chapter (I won't spoil the fun by telling you now!)

Now we turn to some arithmetic systems. Until further notice, the word "number" shall mean positive whole number—one of the numbers 1, 2, 3, 4, ... etc. There are several notations for numbers. One is the familiar base 10 notation, which uses the digits from 0 to 9. Then there is the binary notation which uses just the digits 0 and 1. There is also, what might be called *unary* notation in which a number n is represented by taking a single symbol—say "#" and representing a positive whole number n by a string of #'s of length n. Thus in this notation, 2 would be represented by ##; 3 by ###, and so forth. As an example of elementary formal systems for arithmetical concepts, let's consider some elementary formal systems over the alphabet [#]—the alphabet whose only symbol is "#."

Suppose we wish to represent the set of even numbers. The following set of axioms would obviously work:

E##

Ex → Ex##

Clearly E represents the set of even numbers.

A still simpler system that works contains just one axiom:

Exx

Here is another program that works. In this case "Ox" stands for "x is odd," and "Ex" stands for "x is even."

O#

Ox → Ex#

Ex → Ox#

Now lets consider the relation of *addition* (x + y = z). which we will symbolize "Ax, y, z." In unary notation, such a program is ridiculously simple and has only one formula:

Ax, y, xy

Problem 3. Consider now the familiar base 10 notation which uses the digits 0, 1, 2, 3, 4, 5, 6, 7, 8, 9.

Show how to represent the set of even numbers in base 10 notation [This is quite easy!]

Exercise 1. Represent the addition relation in base 10 notation [This is less easy!]

Some of you are familiar with *binary* notation, in which every whole number is represented by a string of two digits 1 and 0. For any $n + 1$ of these digits, d_0, d_1, ..., d_n, the binary numeral $d_n d_{n-1}, ... d_2 d_1 d_0$ designates the number $2^n d_n + 2^{n-1} d_{n-1} + ... + 2^1 d_1 + 2^0 d_0$. For example the binary numeral 101101 designates the number $2^5 + 0 + 2^3 + 2^2 + 0 + 1$, which is $32 + 0 + 8 + 4 + 0 + 1$, which is 45 (in base 10 notation).

When Von Neumann suggested using binary notation for computers, this revolutionized the whole field of computer science! For a given switch, 1 meant the switch was on and 0 meant the switch was off.

In [Smullyan, 1961] I introduced a variant of binary notation which I called *dyadic* notation and which has certain technical advantages over binary notation. Just as any whole number can be uniquely expressed in binary notation by a string of 1's and 0's, so can any *positive* whole number be uniquely expressed as a string of 1's and 2's. We let D be the alphabet {1, 2}, and we call these two digits *dyadic* digits. Any string $d_n d_{n-1}, ... d_2 d_1 d_0$ of dyadic digits denotes the positive integer

$$2^n d_n + 2^{n-1} d_{n-1} + ... + 2^1 d_1 + 2^0 d_0.$$

Note that the *equation* for evaluating the dyadic numeral is the same as in the case of binary notation that you just saw above. It is just that for any given number in decimal notation, the digits of the dyadic and binary expressions of the number are different. For example the dyadic numeral 1211 denotes $1 \times 2^3 + 2 \times 2^2 + 1 \times 2^1 + 1 \times 2^0$, which is $8 + 8 + 2 + 1$, which is nineteen. Here are the first twenty positive integers in both dyadic and binary notations:

dyadic		binary
1	1	1
2	2	10
3	11	11
4	12	100
5	21	101
6	22	110
7	111	111
8	112	1000
9	121	1001
10	122	1010
11	211	1011
12	212	1100
13	221	1101
14	222	1110
15	1111	1111
16	1112	10000
17	1121	10001
18	1122	10010
19	1211	10011
20	1212	10100

We now consider the alphabet D consisting of the two symbols 1 and 2. Any elementary formal system over D will be called an *elementary dyadic system*, or a *dyadic system*, for short. [In [Smullyan, 1961], this kind of system was called a *dyadic arithmetic*]. A set or relation of numbers will be called *recursively enumerable* (abbreviated r.e.) if it is representable in some dyadic system. [This is actually equivalent to being represented in base n notation for any n, but as mentioned before, dyadic notation has several technical advantages.]

A set or relation of numbers is called *recursive*, or *solvable* if both it and its complement are recursively enumerable.

The study of recursively enumerable and recursive relations is the subject known as *recursion theory*. A fundamental result of this theory which will be proved in this chapter is that there exists a recursively enumerable set that is not recursive. This is closely related to Gödel's incompleteness theorem, as we will see.

Many of the familiar arithmetic operations, such as addition, multiplication, and exponentiation, are recursively enumerable (indeed, recursive) as we will see. To get a solid start on showing these things, we must first establish the following basic results:

By the *successor* of a number x is meant the number immediately following x—i.e. the number x + 1.

Problem 4. (a) Prove that the relation "x is the successor of y" is recursively enumerable.

(b) Now prove that the following relations are recursively enumerable:

(1) The relation $x < y$ (x is less than y)
(2) The relation $x \leq y$ (x is less than or equal to y)
(3) The relation $x = y$ (x equals y)
(4) The relation $x \neq y$ (x is unequal to y)

Functional Relations. A relation $R(x, y)$ is called *functional* if for every number x, there is one and only one y such that $R(x,y)$ holds. Similarly for relations of three or more arguments—e.g. a relation $R(x, y, z)$ is called *functional* if for every x and y, there is one and only one number z such that $R(x, y, z)$ holds:

Problem 5. Prove that if a relation is recursively enumerable and functional, then it is recursive.

Problem 6. Prove that the following relations are recursive

(a) $x + y = z$
(b) $x \times y = z$
(c) $x^y = z$

Gödel Numbering. Consider an ordered alphabet K—i.e. an alphabet whose symbols are arranged in some sequence $a_1, a_2, ..., a_n$. We wish to assign to each string of symbols of K a code number, or a Gödel number in such a way that different strings have distinct Gödel numbers. Several ways of doing this are in the literature, and the way we will do it uses dyadic numerals.

To the symbols $a_1, a_2, ..., a_n$ we respectively assign the numbers 12, 122, ..., 122 ...2 (i.e., in the last case, 1 followed by n occurrences of 2), and for a complex expression, we replace each symbol by its Gödel number—for example we will take the Gödel number of $a_3a_1a_2$ to be 122212122.

This is the Gödel numbering that we will use. It has the great technical advantage that for any pair of expressions X and Y, with respective Gödel numbers x and y, the Gödel number of XY (X followed by Y) is xy (x followed by y). For example if $X = a_2a_3$ and $Y = a_1a_4$, then XY is $a_2a_3a_1a_4$ and its Gödel number is 1221222 (the Gödel number of X) followed by 1212222 (the Gödel number of Y), which is 12212221212222. [In mathematical terminology, this Gödel numbering is an *isomorphism* with respect to concatenation.]

For any K-string X, we will let X_0 stand for it's Gödel number. And for any set W of strings in K, by W_0 we shall mean the set of Gödel numbers of all the elements of W. The following is crucial:

Proposition 2. If W is formally representable then W_0 is recursively enumerable.

Problem 7. Prove Proposition 2, recalling that what Proposition 2 means, in detail, is the following: If a set of K-strings W is formally representable in some elementary formal system over the alphabet K, then the set of the dyadic Gödel numbers of the members of the set W is formally representable over some dyadic system.

Exercise 2. The converse of Proposition 2 is also true—i.e. if W_0 is recursively enumerable then W is formally representable in an elementary formal system over the alphabet K. Prove this.

Exercise 3. Suppose that an alphabet L contains all symbols of the alphabet K (such an L is called an *extension* of K) and that W is a set of strings in K. Prove that W is formally representable over K if and only if W is formally representable over L. [One direction is easy. The other is helped by Proposition 2 and its converse.]

The Universal System (U) We now wish to construct a so-called "universal system" in which we can express all propositions of the form that such and such a number is in such and such a recursively enumerable set (and more generally, that such and such an n-tuple of numbers stands in such and such recursively enumerable relation).

To this end, we must first "transcribe" all elementary dyadic systems into a finite alphabet. Instead of taking our variables and predicates to be individual symbols, as we did before, we now take three symbols v, ', p and define a *transcribed variable* to be any one of the strings v', v", v''', etc. (i.e. the symbol v followed by one or more accents), and a *transcribed predicate* to be a string of one or more p's followed by one or more accents; the number of p's is to indicate the degree of the predicate. We define a *transcribed dyadic system* abbreviated "T.D." to be a system like an elementary dyadic system, except that instead of using individual symbols for variables and predicates, we use transcribed variables and transcribed predicates. We thus have one single 7-sign alphabet K_7—namely the 5 symbols 1, 2, v, ', p together with the comma and implication sign—from which all T.D.'s are constructed. We define "T.D. term," "atomic T.D. formula" "T.D. formula," "T.D. sentence" as we did for dyadic systems, only using "transcribed variable," "transcribed predicate" instead of "variable," "predicate," respectively.

We now construct our universal system (U) as follows: We extend our alphabet K_7 to an alphabet K_9 by adding two new symbols $*$ and $\vdash$. To every T.D. whose axiom schemes are A_1, ..., A_n we associate the string $A_1*A_2*...*A_n$ (or A_1 alone, if $n = 1$), which we call a *base*. The formulas A_1, ..., A_n are called the components of that base. By a *sentence* of (U) we shall mean an expression of the form $B\vdash X$, where B is a base and X is a transcribed *sentence* from the T.D. corresponding to the base. Thus a sentence of (U) is an expression of the form $A_1*...*A_n \vdash X$ (or $A_1\vdash X$) where A_1, ..., A_n are transcribed formulas and X is a transcribed sentence. We call the sentence *true* if and only if X is provable in that transcribed system whose initial formulas are A_1, ..., A_n.

Representability in (U). By a predicate H of (U) of degree 1 (not to be confused with a T. D. predicate— i.e. a predicate of a T. D. system!) we mean an expression of the form $B\vdash P$, where B is a base $A_1*...*A_n$ and P is a T.D. predicate of degree one. Given any number n (in dyadic notation) the expression Hn (which is $B\vdash Pn$) is either true or false, and we say that H represents (in (U)) the set of all numbers n such that $B\vdash Pn$ is true. Thus the predicate $A_1*...*A_n\vdash P$ represents the very same set that is represented by P in the transcribed system whose initial formulas are A_1, ..., A_n. Thus all recursive enumerable sets are representable in (U), and in this sense, (U) is called a *universal* system.

Similarly, for any number n, we define a predicate of degree n of U to be an expression of the form $B\vdash P$, where B is a base and P is a transcribed predicate of degree n, and the predicate $B\vdash P$ represents in (U) the relation on n-tuples of dyadic numerals m_1, ..., m_n such that the n-tuple m_1, ..., m_n is in the relation iff $B \vdash Pm_1, ..., m_n$ is true.

Now comes our main task: We will show that the set of true sentences of (U) is formally representable! We wish to construct an elementary formal system (W) over the alphabet K_9 in which the set T of true sentences of (U) is represented.

To begin with, we must note that the implication sign of (W) is to be distinct from the implication sign of T.D.'s. We could continue to use "→" as the implication sign of (U), but we prefer to use "→" for the implication sign of (W), since it will occur so frequently. For the implication sign of (U), we will use "imp." Similarly, we shall now use the ordinary comma for our punctuation sign of (W), and "com" for the punctuation sign of T. D.'s. For variables of (W) (not to be confused with T. D. variables) we will use the letters x, y, z, w, with or without subscripts. Predicates of (W) (not to be confused either with predicates of (U), or T. D. predicates) should be introduced as needed.

Now for the construction of (W):

Problem 8. Construct an elementary formal system (W) over the alphabet K_9 in which the set T of true sentences of (U) is represented. This construction can be conveniently done by successively representing the following items.

  (a) The set of numbers (dyadic numerals)
  (b) The set of strings of accents
  (c) The set of transcribed variables
  (d) The set of T. D. predicates
  (e) The set of T. D. terms
  (f) The set of *atomic* T. D. formulas
  (g) The set of T. D. formulas
  (h) The set of atomic T. D. sentences
  (i) The set of T. D. sentences
  (j) The relation "x and y are distinct variables"

(k) The relation sub(x, y, z, w)—i.e. "x is any string compounded from dyadic numerals, T. D. variables, T. D. predicates, com, imp; y is a T. D. variable; z is a dyadic numeral and w is the result of substituting z for all occurrences of y in x which are not followed by more accents. [This is the toughest part of the project!]

(l) The relation "x is a T. D. formula and y is the result of substituting numerals for some (but not necessarily all) variables of x. [Such a formula y is called a *partial instance* of x]

(m) The relation "x is a T. D. formula and y is the result of substituting dyadic numerals for *all* the variables of x. [Such a y is called an *instance* of x—sometimes a *total* instance.]

(n) The set of bases of (U)

(o) The relation "x is a component of the base y"

(p) The set T of true sentences of (U).

We have now proved a fundamental result:

Theorem 1. The set T of true sentences of (U) is formally representable—indeed, has been represented in an elementary formal system over K_9.

This, with the result of Problem 8 yields:

Corollary. The set T_0 of Gödel numbers of the true sentences of (U) is recursively enumerable.

The Recursive Unsolvability of (U). A fundamental result of recursion theory is that there exists a recursively enumerable set that is not recursive (its complement is not also recursively enumerable). We will now show that the set T_0 of Gödel numbers of the set T of true sentences of (U) is such a set. We already know that T_0 is recursively enumerable, and so we are to show how its complement $\widetilde{T_0}$ (the set of all numbers *not* in T_0) is not recursively enumerable. To this end we employ a variant of the famous diagonal argument of Gödel.

Consider any sentence X of (U) and any set A of numbers. We shall say that X is a *Gödel sentence* for A if either X is true and its Gödel number is in A, or X is false (not true) and its Gödel number is not in A—in other words, X is true if and only if its Gödel number is in A. [Intuitively, we can think of a Gödel sentence for A as saying, "My Gödel number is in A" If it is true, its Gödel number is in A, as it says. If not true, its Gödel number is not in A]. It is obvious that there cannot be a Gödel sentence for the set $\widetilde{T_0}$, for such a sentence would be true if and only if its Gödel number were in $\widetilde{T_0}$ which would mean that X is true if and only if its Gödel number was not the Gödel number of a true sentence—in other words, X would be true if and only if it was not true, which of course is impossible. Thus there cannot be a Gödel sentence for the set $\widetilde{T_0}$. Therefore, to show that $\widetilde{T_0}$. is not recursively enumerable, it suffices to show that for every recursively enumerable set A, there is a Gödel sentence for A. We will now do that.

For any string X of symbols of K_9, we define its *norm* to be the string XX_0—i.e. X followed by its own (dyadic) Gödel number. We note that for any predicate H of (U), its norm HH_0 is a sentence which is true if and only if the Gödel number H_0 of H lies in the very set represented in (U) by H. Let us say that a number n *satisfies* a predicate H to mean that the number n lies in the set represented by H—in other words the sentence Hn is true. Thus the norm HH_0 of H is a sentence that is true if and only if H is satisfied by its own Gödel number. Now, every string of symbols of K_9 has a norm, hence in particular, so does every number (dyadic numeral). Thus for any number n, its norm is nn_0 (n followed by its own Gödel number). If we order the symbols of K_9 such that 1 is the first symbol and 2 the second, then the Gödel number of 1 is 12 and the Gödel number of 2 is 122, and so, for example, the norm of the number 121 is 1211212212 (121 followed by its Gödel number 1212212).

Problem 9. Is the following true or false? If n is the Gödel number of X, then the norm of n is the Gödel number of the norm of X.

For any number set A, we define $A^{\#}$ as the set of all numbers n whose norm is in A. Thus $n \in A^{\#}$ if and only if $nn_0 \in A$.

Problem 10. Prove that if A is recursively enumerable, so is $A^{\#}$.

Problem 11. Now prove that for any recursively enumerable set A, there is a Gödel sentence for A.

Having proved this, we see that $\widetilde{T_0}$ cannot be recursively enumerable, since that cannot be a Gödel sentence for $\widetilde{T_0}$. Thus we have a fundamental result:

Theorem 2. The set T_0 of Gödel numbers of the true sentences of the universal system (U) is recursively enumerable but not recursive.

Solutions to the Problems of Chapter XVI

1. Suppose S_1 and S_2 are both K-representable. Let (E_1) be an elementary formal system in which some predicate—say P_1—represents S_1. Then given an elementary formal system in which S_2 is represented, if any predicates of the system are also predicates of (E_1), then simply replace them with new symbols, and so we then have an elementary formal system (E_2) in which S_2 is represented by some predicate—say P_2—and no predicate is in both (E_1) and (E_2). Then let (E) consist of all formulas of (E_1) together with all formulas of (E_2), and let the formulas distinguished as axioms in both systems be joined together to form the set of axioms of the new system (E). Since no predicate is common to both (E_1) and (E_2), then P_1 will still represent S_1 in (E) and P_2 will still represent S_2 in (E).

2. Suppose S_1 and S_2 are both K-representable. By Proposition 1, there is a single elementary formal system (E) in which S_1 and S_2 are both representable—say by predicates H_1 and H_2 respectively. To represent the union $S_1 \cup S_2$, take a new predicate P and the two initial formulas

$H_1x \to Px$
$H_2x \to Px$

Then P represents $S_1 \cup S_2$.

To represent the intersection $S_1 \cap S_2$, take a new predicate Q and add the single formula:

$H_1x \to H_2x \to Qx$

Then Q represents $S_1 \cap S_2$

3. In base 10 notation, a number is even if and only if it ends with 0, 2, 4, 6, or 8. And so the set of even numbers is represented by "E" in the elementary formal system whose initial formulas are:

E0

E2

...

E8

Ex0

Ex2

...

Ex8

4. (a) To begin with, we note that the successor of 1 is 2, the successor of 2 is 11 and for any dyadic numeral x, the successor of x1 is x2 and the successor of x2 is y1, where y is the successor of x. Thus the relation "x is the successor of y" is represented by "S" in the dyadic system:

S1, 2

S2, 11

Sx1, x2

$Sx, y \rightarrow Sx2, y1$

(b) Now add to the above axiom schemes the following ones, and the predicates $<, \leq, =, \neq$ will respectively represent the relations x is less than y, x is less than or equal to y, x equals y, x is unequal to y. [For easier readability we will depart from the strict notation and write these predicates between x and y, instead of in front of them—e.g. we wrote "x < y" instead of "< x, y."]

$Sx, y \rightarrow x < y$

$x < y \rightarrow y < z \rightarrow x < z$

$x < y \rightarrow x \leq y$

$x \leq x$

$x \leq y \rightarrow y \leq x \rightarrow x = y$ (Alternatively x = x)

$x < y \rightarrow x \neq y$

$y < x \rightarrow x \neq y$

5. Suppose the relation $R(x, y)$ is recursively enumerable and functional. Given a dyadic system in which R is represented—say by "R"—containing no predicates of

the system of the last problem, add the axioms of the last problem and take a new predicate—say "$\overline{R}$" and add the axiom $Rx, y \to y \neq z \to \overline{R}x, z$. Then "$\overline{R}$" represents the relation "$R(x, y)$ doesn't hold." The proof for relations of more than two arguments is similar.

6. Since these three relations are functional, it suffices to show that they are recursively enumerable.

(a) The addition relation is completely determined by the following two conditions:

(1) $x + 1$ is the successor of x

(2) $x +$ (the successor of y) is the successor of $x + y$. In other words, if y' is the successor of y and $x + y$ is z, and z' is the successor of z, then $x + y' = z'$.

Thus we take the system in which S represents the relation "the successor of x is y" and add the following:

(1) $Sx, x' \to Ax, 1, x'$

(2) $Sy, y' \to Ax, y, z \to Sz, z' \to Ax, y', z'$.

Then A represents the relation "$x + y = z$".

(b) The multiplication relation is completely determined by the following two conditions:

(1) x times 1 is x

(2) x times the successor of y is (x times y) + x.

We thus take the above system for addition and add the following axiom schemes and then M will represent the multiplication relation $Mx, 1, x$.

$$Sy, y_1 \to Mx, y, z \to Az, x, w \to Mx, y_1, w$$

(c) The exponential relation $x^y = z$ is completely determined by the following conditions:

(1) $x^1 = x$

(2) $x^{y+1} = x^y \times x$

We thus add to the above system the following axiom schemes, and E will then represent the exponential relation.

Ex, 1, x

Ex, y, z $\rightarrow$ Mz, x, w $\rightarrow$ Sy, y_1 $\rightarrow$ Ex, y_1, w

7. We are considering an alphabet K, whose symbols are, say, a_1, a_2, ..., a_n. Without loss of generality, we can assume that the symbols 1 and 2 are not symbols of K (if they are, simply replace them by new symbols other than 1 or 2).

W is a set of strings in K and (E) is an elementary formal system over K in which some predicate P represents W. We are to show that the set W_0 of dyadic Gödel numbers of the elements of W is formally representable over the alphabet {1, 2}. We construct a dyadic system (D) (over {1, 2}) with the same variables, predicates and punctuation symbols as in (E), in which P will represent W_0 as follows:

First we take a new predicate G (one that does not occur in (E)) which is to represent the set of dyadic Gödel numbers of all the strings in K and we take for Group I the following axiom schemes:

G12

G122

..

G122...2 (n 2's, where n is the number of symbols of K)

Gx $\rightarrow$ Gy $\rightarrow$ Gxy

Next, for any axiom scheme X of (E), remember that X_0 is the result of replacing each symbol of K that occurs in X by its dyadic Gödel number (e.g. if X = $a_3a_1a_2$, then X_0 = 122212122). Next, let X_1 be the string $Gx_1 \rightarrow Gx_2 \rightarrow$ . .. $\rightarrow Gx_k \rightarrow X_0$, where x_1, ..., x_k are the variables that occur in X_0. Then for each axiom scheme X of (E), we add X_1 to the axiom schemes of Group I, and this set of axiom schemes completes our dyadic system (D). In it, P represents W_0.

8. We introduce the initial formulas of W in groups, first explaining what each newly introduced predicate of W is to represent.

N represents the set of numbers (dyadic numerals)

N1

N2

$Nx \rightarrow Ny \rightarrow Nxy$

Acc represents the set of strings of accents

Acc'

$Accx \rightarrow Accx'$

V represents the set of transcribed variables

$Accx \rightarrow Vvx$

P represents the set of T. D. predicates.

$Accx \rightarrow Ppx$

$Px \rightarrow Ppx$

t represents the set of T. D. terms

$Nx \rightarrow tx$

$Vx \rightarrow tx$

$tx \rightarrow ty \rightarrow txy$

F_0 represents the set of atomic T. D. formulas

$Accx \rightarrow ty \rightarrow F_0pxy$

$F_0x \rightarrow ty \rightarrow F_0px$ com y

F represents the set of T. D. formulas

$F_0x \rightarrow Fx$

$F_0x \rightarrow Fy \rightarrow Fx$ imp y

S_0 represents the set of atomic T. D. sentences

$Accx \rightarrow Ny \rightarrow S_0pxy$

$S_0x \rightarrow Ny \rightarrow S_0px$ com y

S represents the set of T. D. sentences

$S_0x \rightarrow Sx$

$S_0x \rightarrow Sy \rightarrow Sx$ imp y

d represents the relation "x and y are distinct T. D. variables"

$Vx \rightarrow Accy \rightarrow dx, xy$

dx, y → dy, x

S_b represents the relation "x is any string compounded from T. D. variables, T. D. predicates, com, imp, and y is a T. D. variable, z is a dyadic numeral and w is the result of substituting the dyadic numeral z for all occurrences of y in x that are not followed by more accents."

Nx → Vy → Nz → S_bx, y, z, x

Vx → Nz → S_bx, x, z, z

dxy → Nz → S_bx, y, z, x

Px → Vy → Nz → S_bx, y, z, x

Vy → Nz → $S_b$com, y, z, com

Vy → Nz → $S_b$imp, y, z, imp

S_bx, y, z, w → S_bx$_1$, y, z, w$_1$ → $S_b$xx$_1$, y, z, ww$_1$

pt represents the relation "x is a T. D. formula and y is the result of substituting numerals for some (but not necessarily all) variables of x." [Such a formula y is called a *partial* instance of x]

Fx → S_bx, y, z, w → ptx, w

ptx, y → pty, z → ptx, z

ins represents the relation "x is a T. D. formula, and y is the result of substituting numerals for *all* variables of x" (such a formula y is called an *instance* of x)

ptx, y → Sy → insx, y

B represents the set of bases of (U)

Fx → Bx

Bx → Fy → Bx*y

C represents the relation "x is a component of the base y"

Fx → Cx, x

Cx, y → Fz → Cx, y*z

Cx, y → Fz → Cx, z*y

Tr represents the set T of true sentences of (U)

Cx, y → ins x, z → Try⊢z

$\text{Try} \vdash x \rightarrow \text{Try} \vdash x \text{ imp } z \rightarrow S_0 x \rightarrow \text{Try} \vdash z$

9. Of course it is true. We recall that our Gödel numbering is such that for any expression X and Y, the Gödel number $(XY)_0$ of XY is $X_0 Y_0$. Now, let n be the Gödel number of X. The norm of X is Xn, and so its Gödel number is $X_0 n_0$, which is nn_0 (since $X_0 = n$), which is the norm of n. Thus the Gödel number of the norm Xn of X is the norm nn_0 of the Gödel number of X.

10. Suppose A is recursively enumerable. Let (D) be an elementary dyadic system in which A is represented, say by the letter "A" (for easier readability).

Take a new predicate "G" to represent the relation "The Gödel number of the dyadic numeral x is y" and add the following initial formulas:

G1, 12

G2, 122

$Gx, y \rightarrow Gz, w \rightarrow Gxz, yw$

Then take a new predicate N to represent the relation: "The norm of x is y" and add:

$Gx, y \rightarrow Nx, xy$

Finally, take a new predicate B and add:

$Ay \rightarrow Nx, y \rightarrow Bx$

Then B represents the set $A^\#$.

11. Suppose A is recursively enumerable. Then so is $A^\#$ (by the last problem). Let H be a predicate of (U) that represents $A^\#$ in (U). Thus for any number n, the sentence Hn is a true sentence of (U) iff $n \in A^\#$. Let h be the Gödel number of H. Then Hh is true iff $h \in A^\#$, which is true iff $hh_0 \in A$. Thus Hh is true iff $hh_0 \in A$. But hh_0 is the Gödel number of Hh (by the last problem). Thus Hh is a Gödel sentence for A (it is true iff its Gödel number hh_0 is in A).

CHAPTER XVII

INCOMPLETENESS AND UNDECIDABILITY

Numerical Systems. In chapters XIII, XIV, and XIV, we considered systems in which predicates represented or expressed sets of *expressions.* Such systems might aptly be called *syntactic* systems. As mentioned earlier, in the systems considered by Gödel, predicates represented sets, not of expressions, but of extra-linguistic entities such as *numbers.* In one particular system that we will come to later, the entities in question were the natural numbers 0, 1, 2, ..., n, ... Such systems I will call *numerical* systems, which we will turn to now.

Let us now consider such a system—call it (N)—in which there is again a denumerably infinite set of expressions, some of which are classified as *predicates,* some as *sentences,* and in which some sentences are classified as *provable* sentences, some as *refutable* sentences. To each expression X is associated an expression denoted X' or $\overline{X}$ called the *negation* of X such that if X is a sentence, so is its negation, and if X is a predicate, so is its negation. The setup is the same as that of Chapter XIV, but with the following difference: Instead of associating with each pair X, Y of expressions an expression XY, in the system (N) now under consideration, we associate with each expression X and each natural number n an expression denoted X(n) such that if X is a predicate, then X(n) is a sentence.

Following Gödel, to each expression X is assigned a (natural) number called the *Gödel number* of the expression. Distinct expressions have distinct Gödel numbers. It will prove technically convenient to arrange all the expressions in an infinite sequence E_0, E_1, E_2, ..., E_n, ... according to the magnitude of their Gödel numbers. We shall refer to each n as the *index* of E_n. If E_n happens to be a *sentence*, we shall sometimes alternatively denote it as S_n and call n a *sentence number*, and if E_n is a *predicate*, we shall sometimes denote it H_n.

For any number x and y, by x∗y we shall mean the index of $E_x(y)$.

Again we let P be the set of provable sentences in (N) and R the set of refutable sentences in (N). Again we assume that a sentence X is provable iff its negation is refutable, and refutable iff its negation is provable. For any set W of expressions, we let W_0 be the set of indices of all the elements of W. Thus, e.g., P_0 is the set of all n such that E_n is provable, and R_0 is the set of all n such that E_n is refutable. Again we call two sentences *equivalent* if they are either both provable, both refutable, or both undecidable.

In place of the conditions C_1 and C_2 of Chapter XIV, we are now given:

Condition C_1. For each predicate H and each number n, the sentence H'(n) is the negation of H(n).

Condition C_2. To each predicate H is assigned a predicate $H^\#$ such that for all n, the sentence $H^\#(n)$ is equivalent to H(n∗n).

For every number set A, by its complement $\tilde{A}$ is meant the set of all numbers that are not in A. As before, for any set W of *expressions*, by its complement $\widetilde{W}$ is meant the set of all expressions that are not in W.

Problem 1. Given a set W of expressions, $\overline{W_0}$ is the set of all numbers that are not in W_0. On the other hand, $(\widetilde{W})_0$ is the set of indices of all expressions that are not in W. Are the sets $\overline{W_0}$ and $(\widetilde{W})_0$ necessarily the same?

We say a predicate *represents* a number set A if for all numbers n, the sentence H(n) is provable if and only if n is in A.

Given two disjoint sets A and B of numbers we say that a predicate H *weakly separates* A from B if H represents a superset of A that is disjoint from B. We say that H *strongly separates* A from B if H represents some superset A_1 of A, and H' represents some superset B_1 of B (which is disjoint from A_1, if the system is consistent).

For any predicate H and any set W of expressions, we define H_W to be the set of all numbers n such that H(n) is in W. Let us note that for P, the set of provable sentences, the set H_P is simply the number set represented by H.

Problem 2. Prove that for any predicate H and any set W of expressions, the set $H_{\widetilde{W}}$ and the set $\overline{H_W}$ (the complement of H_W) are the same.

We again assume the system (N) is consistent. Again, a sentence is called *undecidable* if it is neither provable nor refutable; otherwise, if the sentence is either provable or refutable, the sentence is said to be *decidable*. The system is called *complete* if every sentence is decidable, and *incomplete* if at least one sentence is undecidable.

Problem 3. Prove the following:

(a) If either P_0 nor R_0 is representable, the system is incomplete.

(b) If H represents some superset of P_0 disjoint from R_0, then H(n) is undecidable for some n.

(c) If some set A is representable and its complement $\tilde{A}$ is not representable, then the system is incomplete.

(d) Neither $\widetilde{P_0}$ nor $\widetilde{R_0}$ is representable in (N).

The Systems (N) and S(N). Associated with the numerical system (N), we define the *syntactic* system S(N) as follows. It is like (N) except in one regard: Instead of associating with each expression X and each number n, a sentence X(n), for each pair of expressions X and Y of S(N), we define XY to be the sentence X(y) of (N), where y is the *index* of Y. Thus for any expression X and any number n, the expression XE_n of the system S(N) is the sentence X(n) of the system (N). It then follows from the given conditions C_1 and C_2 of the system (N) that conditions C_1 and C_2 of Chapter XIV hold for the associated syntactical system S(N).

Problem 4. Prove this vital fact.

As a consequence of the fact that S(N) satisfies conditions C_1 and C_2 of Chapter XIV, it follows that all results of Chapter XIV hold for the syntactic system S(N), and each such result yields a corresponding result for the numeric system (N). Thus the four parts of Problem 3 are simply results of Chapter XIV in a new dress.

Problem 5. Given a predicate H and a set W of expressions, prove that H represents the set W in the system S(N) if and only if H represents the number set W_0 in the system (N).

Gödel Sentences. We shall call a sentence S_n (a sentence whose index is n) a *Gödel sentence* for a number set A to mean that S_n is provable if and only if n is in A.

Exercise 2. For any number set A, let A* be the set of all numbers n such that n*n is in A.

(a) Prove that if A is representable, so is A*.

(b) Prove that if A* is representable then there is a Gödel sentence for A.

(c) Conclude that every representable set has a Gödel sentence.

Problem 6. Suppose that H is a predicate that represents a set W of expressions in S(N). Prove that a sentence is a fixed point of H in the system S(N) if and only if it is a Gödel sentence for W_0 in the system (N).

Remarks. From the above Problems 2 and 3, and the fixed point theorem of Chapter XIV (Theorem F) it follows that in the system (N), every number set representable in (N) has a Gödel sentence. This provides a solution to part (e) of Exercise 2, which of course could also be solved directly from conditions C_1 and C_2 of the system (N).

Problem 7. Which, if either, of the following statements are true?

(a) The system (N) is incomplete if and only if there is a Gödel sentence for P_0.

(b) The system (N) is incomplete if and only if there is a Gödel sentence for R_0.

Definability. We say that a predicate II *defines* a number set A in the system (N) if for every number n:

(a) If n is in A, then H(n) is provable.

(b) If n is not in A, then H(n) is refutable.

We say that A is *definable* if some predicate defines it.

Problem 8. Assuming consistency, which of the following statements are true?

(a) If A is representable then A is definable.

(b) If A is definable then A is representable.

(c) If H defines A then H represents A and H' represents the complement $\tilde{A}$ of A.

Problem 9. Prove that if some set is representable but not definable then the system (N) is incomplete.

Problem 10. Prove that no superset of P_0 disjoint from R_0 can be definable in the system (N).

Formal Systems. We now connect the subject of incompleteness with the recursion theory aspects of the last chapter.

A minimal requirement that a system be called *formal* (sometimes *axiomatizible*) is that the following sets and relations be recursive:

1. The set S_0 of indices of the set of sentences.
2. The set of indices of all the predicates.
3. The relation "x is the index of the negation of that expression whose index is y—in other words, $E_x = \overline{E_y}$." We abbreviate this relation: neg(x,y) or x neg y.
4. x diag y (x is the index of the diagonalizer of that predicate whose index is y—in other words, $E_x = E_y^{\#}$).
5. x st y (x is the index of a sentence that is provable at stage y)
6. rp(x,y,z)   (x*y = z)

For the rest of this chapter, we assume that (N) is a consistent formal system, and hence satisfies the conditions 1–6.

Problem 11. Prove that the sets P_0 and R_0 are recursively enumerable.

We recall that for any predicate H and any set W of expressions, H_W is the set of all numbers n such that H(n) is in W.

Problem 12. Prove the following:

(a) If W_0 is recursively enumerable, so is H_W.

(b) If W_0 is recursive, so is H_W.

(c) Every set representable in a formal system is recursively enumerable.

(d) Every set *definable* in a formal system is recursive.

(e) If P_0 is recursive, so is every representable set.

Problem 13. Prove that the following two conditions are equivalent (i.e. each of them implies the other).

(a) For every recursively enumerable set disjoint from P_0 there is a Gödel sentence.

(b) The complement $\widetilde{P_0}$ of P_0 is not recursively enumerable.

Problem 14. Prove that if some recursively enumerable but not recursive set is representable, then the system is incomplete.

Problem 15. Prove that if every recursively enumerable set is representable, then the system is incomplete.

The result of Problem 15 is well known. We will soon prove the stronger result that a sufficient condition for a consistent formal system to be incomplete is that every *recursive* set is representable. But first:

Theorem 1. If a consistent formal system is complete, then the set P_0 is recursive.

To prove this important result, the lemma below is helpful. [We recall that the union $A \cup B$ of sets A and B is the set of all elements that are in A or in B].

Lemma. If A is disjoint from B then $\widetilde{A \cup B} \cup B = \tilde{A}$

Problem 16. Prove the above lemma and Theorem 1.

Undecidability. Unfortunately the word *undecidable* has two very different meanings in mathematical logic, which could possibly cause some confusion. On the one hand, a sentence of a system is called *undecidable* in the system if it is neither provable or refutable in the system. This definition makes no reference to recursion theory.

On the other hand the system as a whole is called *undecidable* if the set P of provable sentences is not solvable—or equivalently, if the set P_0 is not recursive. Thus undecidability for *systems* is a notion of recursion theory.

The following is an important tie-up:

Theorem 2. If a formal system is undecidable, it is also incomplete.

Problem 17. Prove Theorem 2 (hah!)

Next we have:

Theorem 3. If every recursive set is representable then the system is undecidable.

Problem 18. Prove Theorem 3.

From Theorems 3 and 2, immediately follows the promised strengthening of Problem 15:

Theorem 4. If every *recursive* set is representable in a consistent formal system, then the system is incomplete.

Recursive Inseparability. The notion of *recursive inseparability* plays a fundamental role in modern approaches to incompleteness and undecidability. Two disjoint number sets A and B are said to be *recursively separable* if there exists a recursive superset A' of A that is disjoint from B (which of course implies that the complement of A' is a recursive superset of B that is disjoint from A, and thus B is also recursively separable from A. We call a pair (A, B) of sets recursively separable if A is recursively separable from B (or equivalently, if B is recursively separable from A). The pair is called *recursively inseparable* if it is not recursively separable.

Problem 19. Given two disjoint sets A and B, are the following two conditions equivalent?

(1) For any disjoint recursively enumerable supersets A', B' of A, B respectively, there is at least one number outside both A' and B'.

(2) The pair (A, B) is recursively inseparable.

Problem 20. In regards to a system (N), which, if either, of the following conditions implies the other?

(a) The system (N) is undecidable.

(b) The pair (P_0, R_0) is recursively inseparable.

We have shown that if every recursive set is representable in (N), then the system is undecidable. We now have the following result:

Theorem 5. If every recursive set is definable in (N) then the pair (P_0, R_0) is recursively inseparable.

Problem 21. Prove Theorem 5.

Problem 22. In preparation for the next problem, given two sets V and W of expressions, prove:

(a) If $V_0 \subseteq W_0$ then $V \subseteq W$.

(b) If $V \subseteq W$ then for any predicate H, $H_V \subseteq H_W$.

(c) If V is disjoint from W then H_V is disjoint from H_W.

Problem 23. Prove the following result:

Theorem 6. For any recursively inseparable pair (A, B):

(a) If A is weakly separable from B in (N) then the system (N) is undecidable.

(b) If A is strongly separable from B in (N) then the pair (P_0, R_0) is in turn recursively inseparable.

Discussion. There do indeed exist recursively inseparable pairs of recursively enumerable sets. In the various formal systems considered by Gödel—such as Peano Arithmetic, which we will turn to later, all recursive sets are definable, hence the sets P_0, R_0 of those systems are recursively inseparable. The sets P_0, R_0 are also recursively enumerable, since the systems are formal.

A completely different method of obtaining recursively inseparable pairs of recursively enumerable sets uses the universal system (U) of the last chapter—cf. my *Theory of Formal Systems* for details [Smullyan, 1961].

Solutions to the Problems of Chapter XVII

In what follows, we shall use the standard notation "$\in$" (epsilon) to abbreviate "is a member of." Thus for any set S and any object x, the expression "$x \in S$" abbreviates "x is a member of S." We also write "$x \notin S$" to abbreviate "x is not a member of S." For any two sets S_1, S_2, the notation "$S_1 \subseteq S_2$" (read "S_1 is a subset of S_2"), means that every member of S_1 is also a member of S_2. If S_1 is a subset of S_2, then also S_2 is called a superset of S_1.

1. For any number n, $n \in \widetilde{W_0}$ iff n is not in W_0, iff $E_n \notin W$. On the other hand, $n \in (\widetilde{W})_0$ iff $E_n \in \widetilde{W}$, iff $E_n \notin W$. Thus each statement $n \in \widetilde{W_0}$ and $n \in (\widetilde{W})_0$ holds iff $E_n \notin W$. Therefore for every number n, we have $n \in \widetilde{W_0}$ iff $n \in (\widetilde{W})_0$, hence the sets $\widetilde{W_0}$ and $(\widetilde{W})_0$ are the same.

2. For any number n and any set W and any predicate H, since $n \in H_W$ iff $H(n) \in W$, it follows that $n \notin H_W$ iff $H(n) \notin W$. Now, $n \in \overline{H_W}$ iff $n \notin H_W$, iff $H(n) \notin W$, iff $H(n) \in \widetilde{W}$, iff $n \in H_{\widetilde{W}}$. Thus $n \in \overline{H_W}$ iff $n \in H_{\widetilde{W}}$. Since this holds for every number n, it follows that the sets $\overline{H_W}$ and $H_{\widetilde{W}}$ are the same.

3. We could prove all of this from scratch (which perhaps the reader has already done) but all parts of this problem are really consequences of results of Chapter XIV, as the reader will see from the next problem and the comments that follow.

4. Given a predicate H and any number n,

(a) by conditions C_1 of the numeric system (N), the sentence H'(n) of System (N) is the negation of H(n). Now, the sentence H'(n) of System (N) is the sentence $H'E_n$ of system S(N), and the sentence H(n) of (N) is the sentence HE_n of S(N). Therefore the sentence $H'E_n$ of S(N) is the negation of the sentence HE_n. Since any expression X is E_n for some n, then for any expression X, the sentence H'X of S(N) is the negation of HX. Thus condition C_1 of Chapter XIV holds for S(N).

(b) We know that:

(1) The sentence $H^{\#}E_n$ of the system S(N) is the sentence $H^{\#}(n)$ of the system (N).

Next we have:

(2) The sentence $H(n{*}n)$ of (N) is the sentence $H(E_nE_n)$ of S(N). [Reason: Since $n{*}n$ is the index of the expression $E_n(n)$ of (N), and $E_n(n)$ of (N) is the sentence E_nE_n of S(N), then $n{*}n$ is the index of E_nE_n, hence $H(n{*}n)$ is the sentence $H(E_nE_n)$ of S(N)].

Now, the sentence $H^{\#}E_n$ of S(N) is the sentence $H^{\#}(n)$ of (N) (by (1)), which is equivalent to $H(n{*}n)$ (by the given condition C_2), and $H(n{*}n)$ is the sentence $H(E_nE_n)$ of S(N). Thus the sentence $H^{\#}E_n$ of S(N) is equivalent to the sentence $H(E_nE_n)$. Since every expression X is E_n for some n, then for every expression X, the sentence $H^{\#}X$ is equivalent to $H(XX)$. Thus condition C_2 holds for S(N).

5. Suppose H represents W in the system S(N). Then for all numbers n, the sentence $H(n)$ of (N) is provable iff the sentence HE_n of S(N) is provable, iff E_n is in W (since H represents W), iff n is in W_0. Thus for every number n, the sentence $H(n)$ of (N) is provable iff n is in W_0, which means that H represents W_0 in the system (N).

Conversely, suppose that in (N), the predicate H represents W_0. Then for every number n, the sentence HE_n of S(N) is provable iff the sentence $H(n)$ of (N) is provable, iff n is in W_0 (since H represents W_0), iff E_n is in W. Thus for all n, the sentence HE_n of S(N) is provable iff E_n is in W, which means that in S(N), the predicate H represents W.

6. We are given that in the system S(N), the predicate H represents W. Hence also, in the system (N), H represents W_0 (by problem 5).

(a) Suppose S_n is a fixed point of H in the system S(N). Thus S_n is provable iff HS_n is provable. Also HS_n is provable iff H(n) is provable (since HS_n is the sentence H(n)), which in turn is the case iff n is in W_0 (since H represents W_0 in the system N)). Thus S_n is provable iff n ∈ W_0, which means that S_n is a Gödel sentence for W_0.

(b) Conversely, suppose that S_n is a Gödel sentence for W_0. Thus S_n is provable iff n is in W_0, which is true iff S_n is in W, which is true iff HS_n is provable (since H represents W in S(N).

Thus S_n is provable iff HS_n is provable, which means that S_n is a fixed point of H.

7. It is statement (b) that is true. Reasons:

(a) Suppose sentence S_n is a Gödel sentence for R_0. Thus S_n is provable iff n is in R_0, which is the case iff S_n is refutable. Thus S_n is provable iff it is refutable. Since we are assuming the system consistent, then S_n is undecidable.

(b) Conversely, suppose S_n is undecidable. Consider the following two false statements:

(1) S_n is provable

(2) n is in R_0

(1) is false since S_n is undecidable. (2) is false, since S_n is not refutable, hence S_n is not in R, hence n is not in R_0.

Now, to say of two statements X and Y that they are equivalent (X iff Y) is to say that they are either both true or both false. Since (1) and (2) are both false, they are equivalent—S_n is provable iff n is in R_0, which means that S_n is a Gödel sentence for R_0.

8. Statement (a) is not generally true (it is true for systems that are complete (and consistent). Statement (c) is true, and statement (b) obviously follows from (c). Thus we will show that (c) is true.

Suppose H defines A. Thus for any number n,

(1) If n is in A then H(n) is provable.

(2) If n is in $\tilde{A}$ then H'(n) is provable (H(n) is refutable).

To show that H represents A, we must show that the converse of (1) holds—i.e., if H(n) is provable then n is in A. Well, suppose H(n) is provable. If n were in $\tilde{A}$, then by (2), H(n) would be refutable, and the system would be inconsistent. Since we are assuming consistency, then n cannot be in $\tilde{A}$, hence n is in A. This proves that H represents A.

To show that H' represents $\tilde{A}$, it suffices to show that if H'(n) is provable, then n is in $\tilde{A}$ (since by (2), if n is in $\tilde{A}$, then H'(n) is provable). And so suppose that H'(n) is provable. If n were in A then H(n) would also be provable (by (1)), and again the system would be inconsistent. Hence n cannot be in A, hence n is in $\tilde{A}$.

This completes the proof.

Remarks. In the terminology of the literature, H is said to "*completely* represent A" if H represents A and H' represents $\tilde{A}$. Thus (c) of the above problem says that (in a consistent system) if H defines A, then H completely represents A. The converse of this is obvious (clearly if H completely represents A then H defines A). Thus (for a consistent system), definability and complete representability are the same thing. Some authors use "bi-representable" for "completely representable.")

9. Suppose A is representable but not definable. Let H be a predicate that represents A. Then of course H(n) is provable for any number n in A. It cannot be that H'(n) is provable for every n in $\tilde{A}$, for if it were, then H would define A, contrary to hypothesis. Therefore there is at least one number n in $\tilde{A}$ such that H'(n) is not provable. Also H(n) is not provable, since n is not in A. Therefore H(n) is undecidable.

10. To begin with, let us note that if H defines some set A, then H(n) must be decidable for every n, because either

n is in A, in which case H(n) is provable, or n is not in A, in which case H(n) is refutable.

Now, if some predicate H defined a superset A of P_0 disjoint from R_0, we would have the following contradiction: On the one hand, H would also represent A (Problem 8) hence H(n) would be undecidable for some n (by Problem 3 (b)). On the other hand, H(n) would be decidable for every n, as we noted above. Thus H cannot define any superset of P_0 disjoint from R_0.

11. (a) To prove that P_0 is recursively enumerable, we are given that the relation x st y (x is the index of a sentence provable at stage y) is recursive, hence it is recursively enumerable. Now, a sentence S_x is provable iff it is provable at some stage y. We thus take a dyadic system in which the relation x st y is represented—say by "st" and we take a new predicate—say "Pr" and add the axiom stx, y → Prx. In that system, "Pr" represents P_0. Hence P_0 is recursively enumerable.

(b) To show that R_0 is recursively enumerable, a sentence S_x is refutable iff it is the negation of some provable sentence S_y. We are given that the relation x neg y. (E_x is the negation of E_y) is recursive, hence recursively enumerable. Also P_0 is recursively enumerable by (a). We thus take a dyadic system in which P_0 is represented—say by "Pr." and the relation x neg y is represented—say by "Ng." We then take a new predicate—say "Rf" and add the axiom Pry → Ng(x, y) → Rfx. Then "Rf" represents the set R_0.

12. (a) Suppose W_0 is recursively enumerable. Given a predicate H, let h be its index. Now, a number n is in H_W iff H(n) ∈ W, iff h∗n ∈ W_0 (since h∗n is the index of H(n)). We are given that the relation x∗y = z is recursively enumerable, and so we take a dyadic system (D) in which this relation is represented—say by "rp," and the set W_0 is represented—say by "A," and take a new predicate—say "B"

and add the axiom rphx,y $\to$ Ax $\to$ By. Then B will represent the set H_W. Thus H_W is recursively enumerable.

(b) Suppose W_0 is recursive. Then W_0 is of course recursively enumerable. Hence by (a), H_W is recursively enumerable. Since W_0 is recursive, then its complement $\widetilde{W_0}$ is also recursively enumerable. Since $\widetilde{W_0}$ is the set $(\widetilde{W})_0$ (Problem 1), then $(\widetilde{W})_0$ is recursively enumerable. Then by (a) (taking $\widetilde{W}$ for W), the set $H_{\widetilde{W}}$ is recursively enumerable. But $H_{\widetilde{W}} = \widetilde{H_W}$ (Problem 2), hence $\widetilde{H_W}$ is recursively enumerable. Thus H_W and its complement $\widetilde{H_W}$ are both recursively enumerable. Thus H_W is recursive.

(c) We already know that P_0 is recursively enumerable (Problem 11). Any representable set is H_P for some predicate H. Since P_0 is recursively enumerable, it follows that H_P is recursively enumerable by (a).

(d) Suppose W_0 is definable in the system. Let H be a predicate that defines it. Then by Problem 8 (c), H represents W_0 and H' represents the complement $\widetilde{W_0}$ of W_0. Hence W_0 and $\widetilde{W_0}$ are both representable. Therefore W_0 and $\widetilde{W_0}$ are both recursively enumerable by (c). Hence the set W_0 is recursive.

(e) If P_0 is recursive, so is H_P (by (b)). Also H_P is the set represented by H. Thus any representable set is H_P for some predicate H, hence is recursive.

13. To begin with, there cannot be a Gödel sentence for the set $\widetilde{P_0}$, for such a sentence S_n would be provable iff it were not provable (S_n would be provable iff $n \in \widetilde{P_0}$, iff $n \notin P_0$, iff $S_n \notin P$, iff S_n is not provable). Thus there is no Gödel sentence for $\widetilde{P_0}$.

(a) Suppose that for every recursively enumerable set disjoint from P_0 there is a Gödel sentence. The set $\widetilde{P_0}$ is obviously disjoint from P_0, and since there is no Gödel sentence for $\widetilde{P_0}$, it cannot be the case that $\widetilde{P_0}$ is recursively enumerable.

(b) Conversely, suppose $\widetilde{P_0}$ is not recursively enumerable.

Let A be any recursively enumerable set disjoint from P_0. Then A is a subset of $\widetilde{P_0}$. Since $\widetilde{P_0}$ is not recursively enumerable and A is recursively enumerable, then A is not the whole of $\widetilde{P_0}$, hence there is at least one number n in $\widetilde{P_0}$ but not in A. Since $n \in \widetilde{P_0}$, then it is false that $n \in P_0$, hence it is false that S_n is provable. Since n is not in A, then it is false that $n \in A$. Thus both statements S_n is provable, $n \in A$ are false, hence they are equivalent (both true or both false), and so S_n is provable iff $n \in A$. Thus S_n is a Gödel sentence for A.

14. Suppose that some recursively enumerable but not recursive set A is representable in (N). Since A is not recursive then $\widetilde{A}$ is not recursively enumerable, hence not representable in (N) (by (c) of Problem 12). Thus A is a representable set whose complement is not representable, hence the system is incomplete (Problem 3(c)).

15. There are two ways of proving this:

Proof 1. We showed in the last chapter that there exists a recursively enumerable but not recursive set, namely the set of dyadic Gödel numbers of the true sentences of the universal system (U). Therefore if *every* recursively enumerable set is representable, so is this recursively enumerable but not recursive set, hence the system is incomplete by the last problem.

Proof 2. Suppose that every recursively enumerable set is representable. The set P_0 is recursively enumerable (Problem 11), hence P_0 is then representable, hence the system is incomplete (Problem 3).

16. First for the lemma: Suppose A is disjoint from B. To show that $\overline{A \cup B} \cup B = \widetilde{A}$, we show that for every number n, it is in $\overline{A \cup B} \cup B$ iff it is in $\widetilde{A}$.

(a) Suppose $n \in \overline{A \cup B} \cup B$. Then either $n \in \overline{A \cup B}$ or $n \in B$. If $n \in \overline{A \cup B}$ then n is outside both A and B, hence of course outside A, hence in $\widetilde{A}$. If on the other hand $n \in B$, then again n ∈ Ã, since B is disjoint from A. Thus in either case, n ∈ Ã. This proves that every element of $\overline{A \cup B} \cup B$ is in Ã.

(b) Conversely, suppose n ∈ Ã. Thus n is not in A. Now either n is in A ∪ B or it isn't.

Case 1. Suppose n is in A ∪ B. Since n is not in A, then it must be in B. Thus in this case, n ∈ B.

Case 2. Suppose n is not in A ∪ B. Then in this case, n ∈ $\overline{A \cup B}$.

Hence either n ∈ $\overline{A \cup B}$ (Case 2) or n ∈ B (Case 1). Thus n ∈ $\overline{A \cup B} \cup B$. This proves that all elements of Ã are in $\overline{A \cup B} \cup B$.

By (a) and (b), Ã = $\overline{A \cup B} \cup B$.

Now to prove Theorem 1: We are to show that if a consistent formal system (N) is complete, then the set P_0 is recursive. Well suppose (N) is such a system. Since it is formal, the sets P_0 and R_0 are recursively enumerable (Problem 11). Since the system is complete then every sentence is either provable or refutable, which means that every sentence number m is either in P_0 or R_0, hence is in $P_0 \cup R_0$. Of course, every number in $P_0 \cup R_0$ is a sentence number, and therefore $P_0 \cup R_0$ is the set of all sentence numbers. Part of the requirements of a system being formal is that the set of sentence numbers is recursive. Thus the set $P_0 \cup R_0$ is recursive. Thus $\overline{P_0 \cup R_0}$ is recursively enumerable. Also by our assumption of consistency, the sets P_0 and R_0 are disjoint. Now it follows from Problem 2 of Chapter XVI that the union of any two recursively enumerable sets is recursively enumerable. Also the sets P_0 and R_0 are disjoint. Since $\overline{P_0 \cup R_0}$ and R_0 are both recursively enumerable, so is their union $\overline{P_0 \cup R_0} \cup R_0$ but this set is $\tilde{P_0}$ (by the above lemma, taking P_0 for A and R_0 for B). Thus $\tilde{P_0}$ is recursively enumerable. Since P_0 is also recursively enumerable, then P_0 is recursive.

17. I said "hah!" because Theorem 2 is simply a re-statement of Theorem 1! Look, Theorem 1 says that if a formal system is complete, then P_0 is recursive—in other words that the system is then decidable. Hence if a formal system is undecidable, it cannot be complete.

18. Suppose every recursive set is representable. If the system were decidable, P_0 would be recursive, hence $\widetilde{P_0}$ would be recursive, hence representable, contrary to Problem 3, (d). Hence the system is undecidable.

19. Yes, they are equivalent. Statement (1) can be equivalently stated that there is no complementary pair of recursively enumerable supersets of A and B. [By a complementary pair of sets is meant a pair in which each one is the complement of the other]. Well, suppose (1) holds. Then there cannot be a recursive superset A' of A disjoint from B, for if there were, then its complement $\widetilde{A'}$ would be a recursively enumerable superset of B, hence $(A', \widetilde{A'})$ would be complementary pair of supersets of A and B. Thus there cannot be any recursive superset of A disjoint from B, which means that the pair (A, B) is recursively inseparable.

Conversely, suppose (A, B) is recursively inseparable. Then for any disjoint pair (A', B') of recursively enumerable supersets of A, B, if the pair were complementary, then A' would be recursive, hence a recursive superset of A disjoint from B, contrary to the assumption that (A, B) is recursively inseparable. Hence there is no such complementary pair, which means that (1) holds.

20. It is (b) that implies (a). Suppose (P_0, R_0) is recursively inseparable. Then P_0 cannot be recursive, for if it were, then P_0 would be a recursive superset of itself disjoint from R_0, contrary to the assumption that (P_0, R_0) is recursively inseparable. Thus P_0 is not recursive, and so the system is undecidable.

21. Suppose every recursive set is definable in (N). Let A be any superset of P_0 disjoint from R_0. If A were recursive, then it would be definable in (N), which is contrary to Problem 10. Thus there is no recursive superset of P_0 disjoint from R_0, which means that P_0 is recursively separable from R_0.

22. (a) Suppose $V_0 \subseteq W_0$. Then for any expression E_n, if $E_n \in V$ then $n \in V_0$, hence $n \in W_0$ (since $V_0 \subseteq W_0$), hence $E_n \in W$. Thus every expression E_n in V is also in W, which means that V is a subset of W.

(b) Suppose $V \subseteq W$. Then for any number n, if $n \in H_V$, then $H(n) \in V$, hence $H(n) \in W$ (since $V \subseteq W$), hence $n \in H_W$. Thus every number n in H_V is also in H_W, which means that $H_V \subseteq H_W$.

(c) Suppose V is disjoint from W. Then for any number n in H_V the sentence $H(n) \in V$, hence $H(n) \notin W$ (since V is disjoint from W), hence $n \notin H_W$. Thus no number n in H_V is also in H_W, hence H_V is disjoint from H_W.

23. Let (A, B) be a recursively inseparable pair.

(a) Suppose A is weakly separable from B in (N). Let H be a predicate that weakly separates A from B in (N). Then $A \subseteq H_P$ and H_P is disjoint from B. If P_0 were recursive, then H_P would be recursive (by (b) of Problem 12, taking P for W), hence H_P would then be a recursive superset of A disjoint from B, making A recursively separable from B, contrary to the assumption that (A, B) is not recursively separable. Therefore P_0 cannot be recursive—the system is thus undecidable.

(b) Suppose H *strongly* separates A from B. Thus $A \subseteq H_P$ and $B \subseteq H_R$.

If (P_0, R_0) were recursively separable, then (A, B) would also be, which is contrary to the given condition that (A, B) is recursively inseparable. Here is why:

Suppose (P_0, R_0) is recursively separable. Then some recursive superset of P_0 is disjoint from R_0. This superset is W_0 for some set W of expressions. Since $P_0 \subseteq W_0$, then $P \subseteq W$ (by Problem 23 (a)), hence $H_P \subseteq H_W$ (Problem 22, (b)). Since $A \subseteq H_P$ and $H_P \subseteq H_W$, then obviously $A \subseteq H_W$. Also, since W_0 is recursive, so is H_W (Problem 12, (b). Thus:

(1) H_W is a recursive superset of A.

Next, since W_0 is disjoint from R_0, then W is obviously disjoint from R, hence H_W is disjoint from H_R (Problem 22, (c)). Since $B \subseteq H_R$ and H_R is disjoint from H_W, then obviously B is disjoint from H_W. Thus H_W is a recursive superset of A disjoint from B. Hence A is recursively separable from B.

This means that if the pair (P_0, R_0) is recursively separable, so is the pair (A, B). But we are given that (A, B) is not recursively separable, hence (P_0, R_0) cannot be recursively separable—it is recursively inseparable.

This concludes the proof.

CHAPTER XVIII
FIRST-ORDER ARITHMETIC

The incompleteness theorems that we have so far considered are of an extremely general sort—they are of the form that if a mathematical system has such and such properties then it is incomplete. We have not yet considered any specific significant system and shown that it does have those properties. We will later consider such a system—the system known as *Peano Arithmetic*, which plays a significant role in much of modern research, and prove Gödel's result that it is incomplete. The system of Peano Arithmetic is based on the language known as first-order logic, which we consider in this chapter. For now, we deal with only the *semantics* of first-order arithmetic—we consider only the aspects concerned with meaning and truth. The *syntactic* aspects, which deal with provability in a formal system, will be considered only in a later chapter.

The Setup. The alphabet K of first-order arithmetic uses some or all of the following 14 symbols:

0 ' ( ) + × ~ ∧ ∨ ⊃ ∀ ∃ v =

The expressions 0, 0', 0'', 0''', ... are called *numerals*— more specifically *Peano Numerals*, and serve as names for the natural numbers 0, 1, 2, 3, ... Thus the name of the natural number n consists of the symbol 0 followed by n accents.

The accent symbol ' is to be thought of as standing for "successor." Thus 0' is the successor of 0, 0" is the successor of 0', and so forth. For every number n, we let $\bar{n}$ be the numeral that names n—e.g. $\bar{5}$ = 0'''''.

The symbols + and × stand, as usual, for plus and times, the addition and multiplication functions. The symbol = stands, as usual, for equals, the predicate stating equality. Parentheses are used for punctuation.

The symbols ~, ∧, ∨, and ⊃ are called the *logical connectives*, and stand respectively for *not, and, or* and *implies*. For any propositions p and q, the expression p ∨ q is to read "At least one of p, q is true," not as "exactly one of p, q is true." The proposition p ⊃ q is to be read "if p, then q" or as "p implies q." It is to be understood as saying nothing more nor less than it is not the case that p is true and q is false—or equivalently, that either p is false, or p and q are both true.

We use the abbreviation "p ≡ q" for (p ⊃ q) ∧ (q ⊃ p), which means that each of p and q implies the other—in other words, p and q are either both true or both false. We read "p ≡ q" as "p is equivalent to q," or "p iff q."

The symbols ∀ and ∃ are called *quantifiers*. ∀ is the universal quantifier and stands *"for all."* ∃ is the *existential* quantifier and stands for *"there exists."*

We also need infinitely many expressions v_1, v_2, v_3, ... called *variables*, which roughly speaking stand for arbitrary numbers. We wish to stay within a finite alphabet, and so we will take for v_1, v_2, v_3, ... the respective expressions (v'), (v"), (v'''), ... Thus v_n is to consist of v followed by n accents, the v and its accompanying accents all enclosed in parentheses. We shall use the letters x, y, z, w as standing for arbitrary variables.

The notions of *term, atomic formula, formula, sentence, free and bound occurrence of variables, substitution, designation* and *truth* are given by the following conditions:

(1) An expression is called a *term* if its being so is a consequence of the following rules:
  (a) Every numeral $\bar{n}$ and every variable is a term.
  (b) If t_1 and t_2 are terms, so are $(t_1 + t_2)$, $(t_1 \times t_2)$ and t_1'.

(2) By a *constant term* is meant a term in which no variable appears. Each constant term *designates* a unique natural number in accordance with the following rules:
  (a) The numeral $\bar{n}$ designates the number n.
  (b) If term t_1 designates n_1 and term t_2 designates n_2, then $(t_1 + t_2)$ designates the number n_1 plus n_2, and $(t_1 \times t_2)$ designates n_1 times n_2, and t_1' designates n_1 plus 1.

(3) By an *atomic formula* is meant an expression of the form $(t_1 = t_2)$, where t_1 and t_2 are terms. It is called an atomic *sentence* if t_1 and t_2 are constant terms.

(4) An expression is called a *formula* if its being so is a consequence of the following rules:
  (a) Every atomic formula is a formula
  (b) If F is a formula, so is ~F.
  (c) For any formulas F and G, the expressions $(F \wedge G)$, $(F \vee G)$ and $(F \supset G)$ are formulas.
  (d) For any formula F and any variable x, the expressions $\forall xF$ and $\exists xF$ are formulas. The formula $\forall xF$ is called the *universal quantification* of F with respect to x, and the formula $\exists xF$ is called the *existential quantification* of F with respect to x.

Note: In displaying formulas, outermost parentheses can be deleted, if no ambiguity results.

(5) An occurrence of a variable x in a formula is called *free* or *bound* (not free) if its being so is a consequence of the following (unfortunately messy) conditions:

 (a) In an atomic formula, all occurrences of x are free.

 (b) The free occurrences of x in ~F are those of F. The free occurrences of x in (F ∧ G), (F ∨ G), (F ⊃ G) are those of F and those of G.

 (d) All occurrences of x in ∀xF are bound, and all occurrences of x in ∃xF are bound

 (e) For any variable y distinct from x, the free occurrences of y in ∀xF are those of F, and the free occurrences of y in ∃xF are those of F.

(6) A formula is called a *sentence* if no variable has any free occurrence in it. Sentences are also called *closed formulas*. Formulas that are not closed are also called *open*.

Some Notations and Abbreviations

The variables in alphabetical order are (v'), (v''), (v'''), ... etc. Any formula F with just one free variable x, is also written F(x), and for any numeral $\bar{n}$, we can write $F(\bar{n})$ for the result of replacing every free occurrence of x in F with $\bar{n}$. For any formula F with just two free variables x and y, we can write F as F(x, y), where x is alphabetically prior to y, and for any numerals $\bar{n}$ and $\bar{m}$; we write $F(\bar{n}, \bar{m})$ for the result of simultaneously replacing all free occurrences of x in F with $\bar{n}$ and all free occurrences of y in F with $\bar{m}$. Similarly with formulas with three or more free variables.

For any formulas F_1, F_2, F_3, we can write $F_1 \wedge F_2 \wedge F_3$ for $((F_1 \wedge F_2) \wedge F_3)$. Similar with ∨ in place of ∧. Similarly for more than three formulas.

(7) Now for the critical notion of *truth*. A sentence is called *true* if and only if its being so is a consequence of the following conditions:

(a) For any constant terms t_1 and t_2, the sentence $(t_1 = t_2)$ is true if and only if t_1 and t_2 designate the same number.

(b) A sentence $\sim$S (the negation of S) is true iff S is not true. For sentences S_1 and S_2, the sentence $(S_1 \wedge S_2)$ is true iff S_1 and S_2 are both true. The sentence $(S_1 \vee S_2)$ is true iff at least one of the two sentences S_1, S_2 is true. The sentence $(S_1 \supset S_2)$ is true iff either S_1 is false (not true) or S_2 is true— in other words, iff it is not the case that S_1 is true and S_2 is false.

(c) For any formula F with just one free variable x, the sentence $\forall xF$ is true iff $F(\overline{n})$ is true for *every* number n, and $\exists xF$ is true iff $F(\overline{n})$ is true for at least one number n.

Two sentences are called *equivalent* if they are either both true or both false. We shall abbreviate "S_1 is equivalent to S_2" by "S_1 equ S_2".

Exercise 1. Prove the following equivalences:

(1) $(S_1 \wedge S_2)$ equ $\sim(\sim S_1 \vee \sim S_2)$
(2) $(S_1 \vee S_2)$ equ $\sim(\sim S_1 \wedge \sim S_2)$
(3) $(S_1 \supset S_2)$ equ $\sim(S_1 \wedge \sim S_2)$
(4) $(S_1 \supset S_2)$ equ $\sim S_1 \vee S_2$
(5) $(S_1 \vee S_2)$ equ $\sim S_1 \supset S_2$
(6) $(S_1 \wedge S_2)$ equ $\sim(S_1 \supset \sim S_2)$
(7) $(S_1 \wedge S_2) \supset S_3$ equ $(S_1 \supset S_2) \supset S_3$
(8) $\forall xF$ equ $\sim(\exists x\sim F)$
(9) $\exists xF$ equ $\sim(\forall x\sim F)$

Arithmetic Sets and Relations. A formula F(x) with just one free variable is said to *express* the set of all numbers n such that $F(\bar{n})$ is true. A formula $F(x, y)$ with just two free variables is said to express the relation $R(x, y)$ if for all numbers n and m, the sentence $F(\bar{n}, \bar{m})$ is true if and only if $R(n, m)$ holds. Similarly with formulas with three or more free variables.

A set or relation is called *arithmetic* if there is a formula that expresses it. Note that the adjective "arithmetic" defined here is pronounced with an accent on the third syllable "me", i.e. a-rith-me'-tic, as opposed to the pronunciation of the arithmetic students study in elementary school, which has an accent on the second syllable "rith," i.e. a-rith'-me-tic.

Problem 1. Show the following to be arithmetic.

Expression	Meaning
(a) pos(x)	x is positive
(b) x < y	x is less than y
(c) x ≤ y	x is less than or equal to y
(d) x div y	x divides y evenly
(e) x pdiv y	x properly divides y—i.e. x divides y and x ≠ 1 and x ≠ y
(f) prm(x)	x is a prime number

Admissible Gödel Numberings. For any expressions X and Y, as usual, by XY is meant X followed by Y. XY is called the *concatenation* of X with Y. The operation of concatenation is associative—i.e. for any expressions X, Y and Z, the expression (XY)Z is the same as X(YZ)—that is, XY followed by Z is the same expression as X followed by YZ.

We let g be a Gödel numbering of all expressions of first-order arithmetic. Thus g assigns to every expression a unique positive integer, and distinct expressions have distinct Gödel numbers. We will also call the Gödel number of an expression the g-number of the expression, and for any g-number n, we let E_n be the expression whose g-number is n.

We now consider an operation that assigns to each number x and each number y (in that order) a number denoted x^y. We shall call this operation a *concatenation operation* if for all g-numbers x and y, the number x^y is the g-number of $E_x E_y$. And now we shall call the Gödel numbering *arithmetically admissible*, or just *admissible* for short, if the relation x^y = z is arithmetic. Until further notice, we shall assume the concatenation operation is admissible.

Exercise 2. Using mathematical induction on n, prove that for any g-numbers $x_1, ..., x_n$.

(a) x_1^x_2^...^x_n is the g-number of $E_{x_1} E_{x_2} ... E_{x_n}$.

(b) The relation x_1^x_2^...^x_n = y is arithmetic.

Here is our overall plan: In this chapter we prove Tarski's theorem for first-order arithmetic—namely that for any admissible Gödel numbering, the set of Gödel numbers of the *true* sentences is not arithmetic. Now, the dyadic Gödel numbering that was defined in Chapter XVI and which we are going to use here is easily shown to be admissible, hence in particular, the set of dyadic Gödel numbers of the true sentences is not arithmetic.

In the next chapter we show that arithmetic truth is not formalizable—that is, there exists no elementary formal system in which the set of true sentences of first-order arithmetic is representable. We do this by showing that given any elementary formal system, the set of dyadic Gödel numbers of any representable set of the system *is* arithmetic, hence by Tarski's theorem, cannot be the set of Gödel numbers of the true sentences of first-order arithmetic.

In the final chapter we consider a formal axiom system for elementary arithmetic—the system known as *Peano Arithmetic*. We show that the set of provable sentences of Peano Arithmetic *is* formally representable—indeed, we construct an elementary formal system in which that set is representable. Since the set of provable sentences of Peano Arithmetic is formally representable and the set of true sentences is not, then the two sets don't coincide. Under the reasonable assumption that all provable sentences of Peano Arithmetic are true, it follows that provables are only a *proper* subset of the trues—some sentence is true but not provable in Peano Arithmetic. Indeed, from the elementary formal system constructed, we can actually *exhibit* a true but not provable sentence. That is our overall plan.

Coming back to this chapter, the main technical hurdle in the proof of Tarski's theorem is this: For any number n, the numeral $\bar{n}$ designating n, like any other expression, has a Gödel number $g(\bar{n})$. What is to be shown is that for any *admissible* Gödel numbering g, the relation $g(\bar{n}) = m$ (as a relation between n and m) is an arithmetic relation. This is no mean task!

Dyadic Concatenation. To this end, we turn to the dyadic notation for positive integers which we introduced in the last chapter. The reason that we have called the numbers 0, 0', 0", 0''' ... *Peano* numerals is to avoid possible confusion with *dyadic* numerals, which are strings of 1's and 2's.

For any numbers n and m, by n∗m we shall mean the number expressed in dyadic notation by the dyadic numeral for n, followed by the dyadic numeral for m. [For example 3∗5 = 29, since the dyadic numerals for 3, 5, and 29 are respectively 11, 21, and 1121]. We now need to show that the relation x∗y = z is arithmetic.

By the *length* of a number in dyadic notation, is meant the number of occurrences of the symbols 1 and 2 in n. [For example, the length of 212 is 3; the length of 11211 is 5]. We let len(n) be the length of the dyadic numeral n. Now, what is x∗y in terms of plus and times? Well, $x*y = x \times 2^{len(y)} + y$ [i.e., x∗y equals x times 2 to the power of len(y), plus y]. We must first show that the relation $x = 2^{len(y)}$ is arithmetic. Now, for any number r, the smallest dyadic numeral of length r consists of a string of 1's of length r, and this number is $2^r - 1$. The largest dyadic numeral of length r consists of a string of 2's of length r, which is twice the smallest dyadic numeral of length r, hence is $2 \times (2r - 1)$. Thus, a number y is expressed in dyadic notation has length r iff y lies between $2^r - 1$ and $2 \times (2^r - 1)$. From this follows that $x = 2^{len(y)}$ iff the following two conditions hold:

C_1. $x - 1 \leq y \leq 2 \times (x - 1)$

C_2. x is a power of 2.

Problem 2. Prove the $x = 2^{len(y)}$ iff conditions C_1 and C_2 both hold.

Condition C_1 is obviously arithmetic because it can be written

$\exists z((z + 1 = x) \wedge (z \leq y) \wedge (y \leq 2 \times z))$,

but what about C_2? How can we express the condition that x is a power of 2 in terms of plus and times, since we don't yet have the exponential relation? Well, we can use a very clever idea due to the logician John Myhill: Since 2 is a prime number, x is a power of 2 iff every divisor of x other than 1 is itself divisible by 2. Thus we have:

$Pow_2(x)$ [x is a power of 2]: iff $\forall y((y \text{ div } x \wedge y \neq 1) \supset 2 \text{ div } y)$

Now that we have seen that the conditions C_1 and C_2 are arithmetic, we have proved that the relation $x = 2^{len(y)}$ is arithmetic (it is expressed by the concatenation of the formulas we have found for C_1 and C_2), and so it follows that the desired relation x∗y = z is arithmetic.

$x*y = z \text{ iff } \exists w((w = 2^{len(y)}) \wedge (x \times w) + y = z))$

The operation of dyadic concatenation is associative—i.e. $(x{*}y){*}z = x{*}(y{*}z)$, and so parenthesis here are not necessary. We can write $x{*}y{*}z$ to mean either order of the operations $(x{*}y){*}z$ or $x{*}(y{*}z)$.

Problem 3. (a) Show that the relation $x_1{*}x_2{*}x_3 = y$ is arithmetic.

(c) Using mathematical induction, show that for any $n \geq 2$ the relation $x_1{*}x_2{*} \ldots {*}x_n = y$ is arithmetic.

Note: From now on, to reduce clutter, we will sometimes write xy for $x{*}y$ (thus xy does not mean "x times y" as it usually does—for that, we will write $x \times y$).

We are still in the process of proving that for any admissible Gödel numbering g, the relation $g(\overline{x}) = y$ is an arithmetic relation between x and y. To this end, we first prove an important lemma due to Quine [Quine, 1946].

Ordered Pairs. By an ordered pair (a, b) of members of a set W we mean a set whose only members are a and b, together with a designation as to which member should be regarded as the first element of the ordered pair and which the second. The first one is written to the left.

Now for the key lemma:

Lemma K. (Quine's lemma). There exists an arithmetic relation $K(x, y, z)$ such that for any finite set S of ordered pairs of positive integers, there is a number z such that for all numbers x and y, the relation $K(x, y, z)$ holds if and only if (x, y) is one of the ordered pairs in S.

To prove this lemma, several preliminaries are in order.

We say that x *begins* y (in symbols xBy) if the dyadic numeral for x begins the dyadic numeral for y—for example, 121 begins 1212212. We say that x *ends* y if the dyadic numeral for x ends the dyadic numeral for y—for example 212 ends 1121212. We say that x is *part of* y (xPy) if the dyadic numeral for x is part of the dyadic numeral for y—for example, 21 is part of 11212. Also if x = y, or x begins y, or x ends y, then x is regarded as part of y. If x begins y and x ≠ y, we shall say that x *properly* begins y. By an *initial segment* of y is meant any x that begins y—if x ≠ y, then we say that x is a *proper* initial segment of y. By an *end* segment of y is meant any x that ends y.

Problem 4. (a) Show that the relations xBy, xEy, xPy are arithmetic.

(b) Show that for any n greater than or equal to 1, the relation $x_1 x_2 \ldots x_n Py$ is arithmetic.

We write x$\overline{P}$y for "x is not part of y." The reader should find it profitable to try the following exercises:

Exercise 3. (a) Show that if x begins y and y begins x then x = y.

(b) Suppose x properly begins yz. Show that either x begins y or there is some proper initial segment z_1 of z such that $x = yz_1$.

More specifically, in order to prove (b), show:

(1) If x is shorter than y, then x properly begins y.

(2) If x is the same length as y, then x = y.

(3) If x is longer than y, then $x = yz_1$ for some proper initial segment z_1 of z.

Following Quine, by a *frame* we shall mean a number of the form 2t2, where t is a string of 1's. We shall say that x is a *maximal* frame of y if x is a frame and x is part of y and x is as long as any frame that is part of y.

Problem 5. Prove the following to be arithmetic:

(a) ones(x) (x is a string of 1's)

(b) fr(x) (x is a frame)

(c) x lf y (x and y are frames and x is longer than y)

(d) x lg y (x is a frame that is longer than any frame that is part of y)

(e) x max y (x is a maximal frame of y)

Now let us consider a finite set S of ordered pairs of positive integers. Arrange the pairs in some sequence

(a_1, b_1), (a_2, b_2), ..., (a_n, b_n). Call this sequence "Θ." Let f be a frame whose string of 1s is longer than any string of 1's that is part of any of the numbers a_1, ..., a_n, b_1, ..., b_n, and let z be the number $ffa_1fb_1ffa_2fb_2ff....ffa_nfb_nff$.

Let us note the following facts about z:

Fact 1. Since the string of 1's in f is longer than the string of 1's in any of the numbers a_1, a_2, ..., a_n, b_1, b_2, ... b_n, f must be a maximal frame of z. Moreover, if g is any maximal frame of z, then g must be the f chosen to construct z.

Fact 2. For any pair of numbers x and y, the pair (x, y) is in S iff $ffxfyff$ is part of z.

Since the reltions x max y, lg x, and xPy are arithmetic, so is the relation defined by

$$\exists\ w(w\ max\ z \wedge w\ lg\ x \wedge w\ lg\ y \wedge wwxwywwPz)$$

(as a relation between x, y and z). Call this relation $K(x, y, z)$. We must show that the arithmetic relation $K(x, y, z)$ does the job it will need to do.

Problem 6. Prove that for any finite set S of ordered pairs of positive integers, there is a number z such that for any numbers x and y, $K(x, y, z)$ holds if and only if (x, y) is one of the pairs in S.

Having solved Problem 6, Lemma K is now proved. The following is an easy corollary:

Theorem K_0. There is an arithmetic relation $K_0(x, y, z)$ such that for any finite set of ordered pairs of *natural* numbers (positive integers or zero), there is a number z such that for any natural numbers x and y, the pair (x, y) is in S iff $K(x, y, z)$ holds.

Problem 7. Prove Theorem K_0.

Problem 8. Prove that for any number z, there are only finitely many pairs (x, y) such that $K_0(x, y, z)$.

Now that we have Theorem K_0, we can prove the desired result that for any admissible Gödel numbering, the relation $g(\overline{x}) = y$ is arithmetic.

We recall that by x^y is meant the g-number of $E_x E_y$. Now suppose g is admissible. Let a be the g-number of $\overline{0}$ and let b be the g-number of the accent. Then $g(\overline{0}) = a$, $g(\overline{1}) = g(\overline{0})^{\wedge}b$, $g(\overline{2}) = g(\overline{1})^{\wedge}b$, ... $g(\overline{n+1}) = g(\overline{n})^{\wedge}b$, etc.

Let us temporarily call a set S of ordered pairs of natural numbers a *special* set if for every pair (x, y) in S, either $x = 0$ and $y = a$, or there is a pair, (x_1, y_1) in S such that $x = x_1 + 1$ and $y = y_1^{\wedge}b$.

Problem 9. (a) Using mathematical indirection, prove that if S is a special set then for every pair (x, y) in S, $g(\overline{x}) = y$.

(b) Now show that for all x and y, $g(\overline{x}) = y$ if and only if the pair (x, y) belongs to at least one special set.

(c) Now, using Theorem K_0, prove that the relation $g(\overline{x}) = y$ is arithmetic.

Tarski's Theorem

Now that we have proved that for any admissible Gödel numbering g, the relation $g(\overline{x}) = y$ is arithmetic, we are ready to prove that for any admissible Gödel numbering g, the set of g-numbers of the *true* sentences of first-order arithmetic is not arithmetic—Tarski's theorem.

Let us take one particular variable—say v_1, and let $F(v_1)$ be a formula with v_1 as its only free variable. For any number n, we recall that $F(\overline{n})$ is the result of substituting the numeral $\overline{n}$ for all free occurrences of v_1 in F. Let $\varphi(n)$ be the g-number of $F(\overline{n})$. The relation $\varphi(n) = m$ (as a relation between n and m) is indeed arithmetic, as Gödel has shown, but to prove this is a complicated matter, which involves having to arithmetize the operation of substitution. Fortunately we can avoid this by using a clever idea due to Alfred Tarski [Tarski, 1953]. Informally the idea is this: To say that a given property P holds for a given number n is equivalent to saying that for any number x equal to n, the property P holds for x. Formally, for any formula $F(v_1)$ with v_1 as the only free variable, $F(\bar{n})$ is equivalent to the sentence $\forall v_1((v_1 = \bar{n}) \supset F(v_1))$. It is also equivalent to $\exists v_1((v_1 = \bar{n} \wedge F(v_1))$. The whole point now is this: Let $\psi(\bar{n})$ be the g-number of the sentence $\forall v_1((v_1 = \bar{n} \supset F(v_1))$. It is a relatively simple matter to show that the relation $\psi(n) = m$ (as a relation between n and m) is arithmetic— far more simple than showing that the relation $\varphi(n) = m$ is arithmetic!

To simplify matters further let us note that for *any* expression E, whether a formula or not, the expression $\forall v_1((v_1 = \bar{n}) \supset E)$ is a well defined expression, though a meaningless one if E is not a formula, and we shall write $E[\bar{n}]$ (notice the *square* brackets!] as an abbreviation of the expression $\forall v_1((v_1 = \bar{n}) \supset E)$. If E is a formula, then $E(\bar{n})$ is a formula, but not necessarily a sentence. But if E is a formula with v_1 as the only free variable, then $E[\bar{n}]$ is equivalent to $E(\bar{n})$ (round brackets). To repeat, if $F(v_1)$ is a formula with v_1 as the only free variable then $F(\bar{n})$ (round brackets) is the result of substituting $\bar{n}$ for all free occurrences of v_1 in $F(v_1)$, whereas $F[\bar{n}]$ (square brackets) is the sentence $\forall v_1((v_1 = \bar{n}) \supset F(v_1))$, and the two sentences $F(v_1)$ and $F[v_1]$ are equivalent.

Given an expression E_x with g-number x, and a number y, what is the g-number of $E_x[\bar{y}]$ in terms of x and y? Well, $E_x[\bar{y}]$ is the expression $\forall v_1((v_1 = \bar{y}) \supset E_x)$. This expression consists of the expression $\forall v_1(v_1 =$ followed by $\bar{y}$ followed by the implication symbol $\supset$ followed by E_x followed by a right parenthesis. Let a be the g-number of $\forall v_1(v_1 =$ let b be the g-number of the implication sign "$\supset$," and let c be the g-number of the right parenthesis. Of course $g(\bar{y})$ is the g-number of $\bar{y}$ and x is the g-number of E_x. Thus the g-number of $\forall v_1(v_1 = \bar{y} \supset E_x)$ is $a{\wedge}g(\bar{y}){\wedge}b{\wedge}x{\wedge}c$. [Of course, for the *dyadic* Gödel numbering, we could write down the numbers a, b, c explicitly]. We now let r(x, y) be the number $a^\wedge g(\bar{y})^\wedge b^\wedge x^\wedge c$, and so, for any expression E_x and any number y, the number r(x, y) is the g-number of $E_x[\bar{y}]$.

Problem 10. Assuming g is an admissible Gödel numbering, prove that the relation r(x, y) = z is arithmetic.

We continue to assume that g is admissible. For any set A of numbers, we define $A^\#$ to be the set of all numbers x such that r(x, x) is in A.

Problem 11. Prove that if A is arithmetic, so is $A^\#$.

We define a sentence S_x (a sentence whose g-number is x) to be a *Gödel sentence* for A if it is the case that S_x is true iff $x \in A$.

Problem 12. Prove that for every arithmetic set A there is a Gödel sentence for A.

A sentence S_n with Gödel number n is said to be a (semantic) *fixed point* of a formula F(x) iff S_n is true iff $F(\bar{n})$ is true.

Problem 13. Suppose a formula F(x) expresses a set A. Which, if either of the following statements are true?

(a) If S_n is a fixed point of F(x), then S_n is a Gödel sentence for A.

(b) If S_n is a Gödel sentence for A, then S_n is a fixed point of F(x).

Problem 14. Given a set A of numbers, which, if either of the following statements implies the other?

(a) Given any arithmetic subset B of A, there is a number n in A but not in B.

(b) The set A is not arithmetic.

For any set W of expressions, we are letting W_0 be the set of g-numbers of all the elements of W. We let T be the set of true sentences of first-order arithmetic and T_0 the set of its g-numbers.

At last, we now have:

Theorem T [*Tarski's theorem,* Tarski, 1953]. Given an admissible Gödel numbering g, for any set W of expressions that contains no false sentences, if the set W_0 of g-numbers of the elements of W is arithmetic, then there is a true sentence that is not in W. In particular, the set T_0 of g-numbers of the true sentences is not arithmetic.

Problem 15. Prove Theorem T.

Solutions to the Problems of Chapter XVIII

1. (a) pos(x) iff $x \neq 0$
   (b) $x < y$ iff $\exists z(\text{pos}(z) \wedge x + z = y)$
   (c) $x \leq y$ iff $\exists z(x + z = y)$
   (d) x div y iff $\exists z(x \times z = y)$
   (e) x pdiv y iff $x \text{ div } y \wedge x \neq 1 \wedge x \neq y$
   (f) prm (x) iff $\sim\exists z(z \text{ pdiv } x)$

2. (a) Suppose $x = 2^{\text{len}(y)}$. Let $r = \text{len}(y)$. Then $x = 2^r$. Since y has length r when expressed in dyadic notation, then $2^r - 1 \leq y \leq 2 \times (2^r - 1)$, and since $2^r = x$, then $x - 1 \leq y \leq 2 \times (x - 1)$. Thus C_1 holds. And of course C_2 holds.

   (b) Conversely, suppose C_1 and C_2 both hold for x and y. Since C_2 holds, then $x = 2^r$ for some r. And by C_1, $x - 1 \leq y \leq 2 \times (x - 1)$. Thus $2^r - 1 \leq y \leq 2 \times (2^r - 1)$, (since $x = 2^r$), which means that the dyadic numeral y is of length r. Thus $r = \text{len}(y)$, so that $x = 2^{\text{len}(y)}$.

3. (a) $x_1 * x_2 * x_3 = y$ iff $\exists z(x_1 * x_2 = z \wedge z * x_3 = y)$

   (b) We start the induction with $n = 2$. We already know that the relation $x_1 * x_2 = y$ is arithmetic. Now suppose $n \geq 2$ and that we know that n is such that the relation $x_1 * x_2 * x_3 * \ldots * x_n = y$ is arithmetic. We must show that $n + 1$ has the same property. Well, $x_1 * x_2 * x_3 * \ldots * x_{n+1} = y$ if and only if
   $\exists z(x_1 * x_2 * x_3 * \ldots * x_n = z \wedge z * x_{n+1} = y)$. This completes the induction.

4. (a) xBy   iff $x = y \vee \exists z(xz = y)$
   xEy   iff $x = y \vee \exists z(zx = y)$
   xPy   iff $xBy \vee xEy \vee \exists z_1 \exists z_2(z_1 x z_2 = y)$
   (b) $x_1 x_2 x_3 \ldots x_n Py$ iff $\exists z(x_1 x_2 x_3 \ldots x_n = z \wedge zPy)$

5. (a) ones(x)   iff $\sim 2Px$
   (b) fr(x)   iff $\exists y(\text{ones}(y) \wedge x = 2y2)$

(c) x lf y   iff $\exists z \exists w(\text{ones}(z) \wedge \text{ones}(w) \wedge y = 2z2$

                         $\wedge x = 2zw2)$

(d) x lg y   iff $\forall z((\text{fr } z \wedge z\text{Py}) \supset x \text{ lf } z)$

(e) x max y   iff $\text{fr } x \wedge x\text{Py} \wedge \forall z((\text{fr } z \wedge z\text{Py} \wedge$

                         $z \neq x) \supset x \text{ lf } z)$

6. We are considering a finite set S of ordered pairs of positive integers, arranged in a sequence Θ, a frame f and a number z, as described before the statement of Problem 6. We show that for that number z, given any pair of positive integers x and y, the relation $K(x, y, z)$ holds if and only if the pair (x, y) is in S.

To reduce clutter, let $M(w, x, y, z)$ be the relation w max z $\wedge$ w lg x $\wedge$ w lg y $\wedge$ wwxwywwPz. Thus $K(x, y, z)$ is the relation $\exists w M(w, x, y, z)$. The relation $M(w, x, y, z)$ is the conjunction of the following three relations.

(1) w max z

(2) w lg x $\wedge$ w lg y

(3) wwxwywwPz

In particular, taking f for w, $M(f, x, y, z)$ is the conjunction of the following three conditions:

(1)' f max z

(2)' f lg x $\wedge$ f lg y

(3)' ffxfyff

(a) In one direction, suppose (x, y) is in z. Then obviously (3)' holds. Also (1)' holds (Fact 1) and we took f such that (2)' holds. Thus (1)', (2)', (3)' all hold, hence their conjunction $M(f, x, y, z)$ holds. Hence there is a number w—namely f—such that $M(w, x, y, z)$ holds. Thus $K(x, y, z)$ [which is $\exists w M(w, x, y, z)$] holds. This proves that if (x, y) is in S, then $K(x, y, z)$ holds.

(b) Going in the other direction, suppose x and y are numbers such that $K(x, y, z)$ holds. Thus there is a number w such that $M(w, x, y, z)$ holds, and therefore (1), (2) and (3) hold. Since (1) holds, then w is a maximal frame that is part of z. Then by Fact 1 and (1), the number w must be the f taken to construct z. So (3)' also holds, and by Fact 2, (x, y) must be in S. This proves that if $K(x, y, z)$ holds, then (x, y) is in S, which completes the proof.

7. Take $K_0(x, y, z)$ to be the relation $K(x + 1, y + 1, z)$. This relation is arithmetic, since it can be written $\exists x_1 \exists y_1 (x + 1 = x_1 \land y + 1 = y_1 \land K(x_1, y_1, z))$.

Now to show that the relation $K_0(x, y, z)$ works.

Let S be a finite set of ordered pairs of *natural* numbers. Let S^+ be the set of all ordered pairs $(x + 1, y + 1)$ of *positive integers* such that the pair (x, y) is in S. Applying Lemma K to the set S^+, there is a number z such that for any natural numbers x and y, the pair $(x + 1, y + 1)$ of positive integers is in S^+ iff $K(x + 1, y + 1, z)$. Thus for any natural numbers x and y, the pair (x, y) is in S iff $(x + 1, y + 1)$ is in S^+, iff $K(x + 1, y + 1, z)$, iff $K_0(x, y, z)$. Thus for any natural numbers x and y, $K_0(x, y, z)$ holds iff (x, y) is in S.

8. We first show this for the relation $K(x, y, z)$. If $K(x, y, z)$ holds then there is a frame w such that, among other things, wwxwywwPz. Then x and y must be less than z, and there are only finitely many pairs (x, y) such that x and y are both less than z. Thus for any given z, there are only finitely many pairs (x, y) such that $K(x, y, z)$. But the number of pairs such that $K_0(x, y, z)$ is the same number, so we are done.

9. (a) Let S be a special set. Let $P(x)$ be the property that for every number y, if the pair (x, y) is in S, then $g(\overline{x}) = y$. We prove that P holds for all natural numbers by mathematical induction. We must first show that $P(0)$ holds, and then that for every number x, if $P(x)$ holds, so does $P(x + 1)$.

To show $P(0)$: Suppose $(0, y) \in S$. Then either $y = a$ or there is a pair (x_1, y_1) in S such that $0 = x_1 + 1$ and $y = y_1 {}^\wedge b$.

The latter alternative is impossible, since 0 cannot be $x + 1$ for any natural number x. Therefore $y = a$. Since $g(\overline{0}) = a$, then $g(\overline{0}) = y$. This proves that for every number y, if $(0, y) \in S$, then $g(\overline{0}) = y$. Thus property P holds for 0.

Next, suppose x is a number such that $P(x)$ holds. We are to show that $P(x + 1)$ holds, in other words that for every number y, if $(x + 1, y)$ is in S, then $g(\overline{x + 1}) = y$.

Well, suppose $(x + 1, y)$ is in S. Then either $x + 1 = 0$ and $y = a$, or there are numbers x_1 and y_1 such that $(x_1, y_1) \in S$ and $x + 1 = x_1 + 1$ and $y = y_1 {}^{\wedge} b$. The first alternative is out, since it is not possible that $x + 1 = 0$. Hence the second alternative holds, namely there are numbers x_1 and y_1 such that $(x_1, y_1) \in S$ and $x + 1 = x_1 + 1$ and $y = y_1 {}^{\wedge} b$. Since $x + 1 = x_1 + 1$, then obviously $x = x_1$, and so $(x, y_1) \in S$ and $y = y_1 {}^{\wedge} b$. Since $(x, y_1) \in S$ and $P(x)$ holds (by the induction hypothesis), then $g(\overline{x}) = y_1$. Thus $y = g(\overline{x}) {}^{\wedge} b$, which is $g(\overline{x + 1})$. Thus $g(\overline{x + 1}) = y$. This proves that if x has the property P, so does $x + 1$. Then, by mathematical induction, it follows that every number x has property P, which means that for every x and y, if $(x, y) \in S$, then $g(\overline{x}) = y$.

Thus, for every special set S and admissible Gödel numbering g, if (x, y) is in S, then $g(\overline{x}) = y$.

(b) Suppose $g(\overline{x}) = y$. Then (x, y) belongs to the set $\{0, g(\overline{0})), (1, g(\overline{1})), \ldots, (x, g(\overline{x}))\}$, which is obviously a special set.

Conversely, suppose (x, y) belongs to some special set S. Then by (a), $g(\overline{x}) = y$, since for *every* pair (x, y) in S, $g(\overline{x}) = y$.

(c) Now comes the interesting part! We consider the relation $K_0(x, y, z)$. Let us call z a *representative* of a finite set S of ordered pairs if for every ordered pair (x, y), the relation $K_0(x, y, z)$ holds iff $(x, y) \in S$. Theorem K_0 is to the effect that every finite set S of ordered pairs has a representative z. Also, every number z is obviously the representative of some finite set S of ordered pairs, namely, the set of all ordered pairs (x, y) such that $K_0(x, y, z)$ holds (which is finite by Problem 8). Of course, for many numbers z, the set S may be empty (for example this will be true for any z that starts with the number 1).

Let us now say that a number z has property Sp—in symbols $Sp(z)$—if the following holds:

$$\forall x \forall y (K_0(x, y, z) \equiv ((x = 0 \land y = a) \lor \text{(continued on next line)}$$
$$\exists x_1 \exists x_2 (K(x_1, x_2, z) \land x = x_1 + 1 \land y = y_1{}^\land b)).$$

[This is read: "For every pair of numbers x and y, $K_0(x, y, z)$ holds if and only if either $x = 0$ and $y = a$, or there exist numbers x_1 and x_2 such that $K_0(x_1, x_2, z)$ holds and $x = x_1 + 1$ and $y = y_1{}^\land b$."] I use the symbol "Sp" to suggest "special," because for any finite set S of ordered pairs, if z is the representative of S, then S is special if and only if $K_0(x, y, z)$ holds. I leave it to the reader to verify that the condition $Sp(z)$ is arithmetic, and therefore that the relation $\exists z(Sp(z) \land K_0(x, y, z))$ is arithmetic. It is a relation between x and y. Now we prove the key fact that this relation holds if and only if $g(\overline{x}) = y$, thus proving that the relation $g(\overline{x}) = y$ is arithmetic.

In one direction, suppose $g(\overline{x}) = y$. Then by (b). the pair (x, y) belongs to some special set S. Let z be a representative of S. Since S is special, then $Sp(z)$ holds. Since z is a representative of S and $(x, y) \in S$, then $K_0(x, y, z)$ holds. Thus $Sp(z) \land K_0(x, y, z)$ holds. Thus, since there is some z such that $Sp(z) \land K_0(x, y, z)$ holds, then $\exists z(Sp(z) \land K_0(x, y, z))$ holds.

Going in the other direction, suppose $\exists z(Sp(z) \land K_0(x, y, z))$ holds. Then there is some number z such that $Sp(z) \land K_0(x, y, z)$ holds. Let S be the set of all ordered pairs (x, y) such that $K_0(x, y, z)$ holds. This set S is finite (by Problem 8), and z is a representative of S. Since $Sp(z) \land K_0(x, y, z)$ holds, then of course $Sp(z)$ and $K_0(x, y, z)$

both hold. Since $Sp(z)$ holds and z is a representative of S, then S is special. Since $K_0(x,y,z)$ holds and z is a representative of S, it follows that the pair $(x,y) \in S$. Thus (x,y) is a member of the special set S, hence by (b), $g(\overline{x}) = y$. Thus if $\exists z(Sp(z) \wedge K_0(x,y,z))$ holds, then $g(\overline{x}) = y$. This completes the proof.

10. Here is where we need the crucial fact that if g is admissible, then the relation $g(\overline{x}) = y$ is arithmetic. Now, $r(x,y) = a^\wedge g(y)^\wedge b^\wedge x^\wedge c$. Then the relation $r(x,y) = z$ can be written $\exists w(g(\overline{y}) = w \wedge z = a^\wedge w^\wedge b^\wedge x^\wedge c)$, which is arithmetic.

11. Suppose A is arithmetic. Since the relation $r(x,y) = z$ is arithmetic, then the relation $r(x,x) = y$ is obviously arithmetic, and since the condition $y \in A$ is arithmetic, then the condition $\exists y(r(x,x) = y \wedge y \in A)$ is arithmetic, and this is the condition that x is in $A^\#$. Thus $A^\#$ is arithmetic.

12. Suppose A is arithmetic. Then by the last problem, $A^\#$ is arithmetic. Let $F(v_1)$ be a formula that expresses $A^\#$. Since for any number n, the sentence $F[\overline{n}]$ is equivalent to $F(n)$ then for any number n, the sentence $F[\overline{n}]$ is true iff $n \in A^\#$. F has some g-number h, thus for all n, the sentence $F_h[\overline{n}]$ is true iff $n \in A^\#$, iff $r(n,n) \in A$. Taking h for n, we see that $F_h[\overline{h}]$ is true iff $r(h,h) \in A$, but $r(h,h)$ is the g-number of $F_h[\overline{h}]$! Thus $F_h[\overline{h}]$ is true iff its g-number $r(h,h)$ is in A, and so $F_h[\overline{h}]$ is a Gödel sentence for A.

13. Both statements are true.

(a) Suppose S_n is a fixed point of $F(x)$. Then S_n is true iff $F(\overline{n})$ is true, iff $n \in A$ (since $F(x)$ expresses A). Thus S_n is true iff $n \in A$, which means that S_n is a Gödel sentence for A.

(b) Conversely, suppose S_n is a Gödel sentence for A. Thus S_n is true iff $n \in A$, but $n \in A$ iff $F(\overline{n})$ is true, and so S_n is true iff $F(\overline{n})$ is true, which means that S_n is a fixed point of $F(x)$.

14. Each one implies the other—the two statements are equivalent.

Suppose (a) holds. Thus for any arithmetic subset B of A, there is a number in A but not in B. Since A is a subset of itself, then if A were arithmetic, there would be a number n in A but not in A, which is impossible. Hence A cannot be arithmetic. Thus (b) holds.

Conversely, suppose (b) holds. Thus A is not arithmetic. Then for any arithmetic subset B of A, B cannot be the whole of A (since B is arithmetic and A is not). Therefore there are numbers in A that are not in B. Thus (a) holds.

15. To begin with, the complement $\tilde{A}$. of any arithmetic set A is arithmetic, because if formula $F(x)$ expresses A, then formula $\sim F(x)$ expresses $\tilde{A}$.

We are given that W contains no false sentence and that W_0 is arithmetic. Then $\widetilde{W_0}$ is arithmetic. Let S_n be a Gödel sentence for $\widetilde{W_0}$. Thus S_n is true iff $n \in \widetilde{W_0}$, iff $n \notin W_0$, iff $S_n \notin W$. Thus S_n is true iff S_n is not in W. This means that either S_n is true and not in W, or S_n is false but in W. The latter altertnative is out, since we are given that W contains no false sentences. Thus S_n is a true sentence which is not in W.

Consider now the case where W is the set T of all true sentences. Obviously no false sentence is in T. If T_0 were arithmetic, then there would be a sentence that would be true but not in T, which is impossible. Therefore T_0 cannot be arithmetic.

CHAPTER XIX
ARITHMETIC TRUTH IS NOT FORMALIZABLE

In this chapter we show that if W is any set of expressions that is representable in some elementary formal system, then the set W_0 of dyadic Gödel numbers of the elements of W is an arithmetic set, i.e. a set expressible by a formula of first-order arithmetic.

This has many important ramifications. For one thing it shows that the set of true sentences of first-order arithmetic is not representable in any elementary formal system, since by the results of the last chapter, the set of dyadic Gödel numbers of the true sentences of first-order arithmetic is *not* arithmetic (Tarski's theorem). Thus, given any axiom system for first-order arithmetic, if the system is *correct*, in the sense that all sentences provable in the system are true, and if the set of provable sentences is formally representable, then there must be a true sentence that is not provable in the system.

Arithmetization of Elementary Formal Systems

Our arithmetization of elementary formal systems is wholly constructive, in the sense that given an elementary formal system in which a set W of expressions is represented, we will see that not only is the set W_0 of its dyadic Gödel numbers an arithmetic set, but we can actually *exhibit* a formula that expresses the set W_0. In the next chapter we consider the well-known axiom system for first-order arithmetic known as *Peano Arithmetic* and we construct an elementary formal system in which the set W of provable sentences of Peano Arithmetic is represented. We can then find a formula—call it "$P(x)$"—that expresses the set W_0. Then the diagonalization of the diagonalizer of $P(x)$ is a sentence that is true but not provable in Peano Arithmetic, hence undecidable in Peano Arithmetic, thus proving Gödel's theorem that Peano Arithmetic is incomplete [Gödel, 1931].

We consider now an elementary formal system (E) over an alphabet K. Let L be the full alphabet of (E). Arrange the symbols of L so that the symbols k_1 ... k_n of K are at the beginning. Thus the alphabet L is ordered in some way like this: k_1, ..., k_n, l_{n+1}, ..., l_m. Now consider the dyadic Gödel numbering of all strings of symbols of L. We now aim to show:

(1) The set of dyadic Gödel numbers of the provable sentences of (E) is arithmetic.

(2) For any set W of expressions constructed from the symbols of K, the set W_0 of dyadic Gödel numbers of the expressions in W is arithmetic.

For any positive integer i we let g_i consist of a single 1 followed by a string of 2's of length i (e.g. $g_3 = 1222$). We let G_n be the set of all strings compounded by concatenation from the g_i's with n or less 2's in its string, where n is the number of symbols in the alphabet K (the symbols in the alphabet K are used to construct constants). Thus G_n is the set of Gödel numbers of *concatenations of the individual symbols* (i.e. the Gödel numbers of the constants) from the distinguished alphabet K of our elementary formal system (E).

Problem 1. Prove that the set G_n is arithmetic.

Substitution Lemma. Assume an elementary formal system (E) with alphabet L of symbols a_1, ... a_m and a distinguished sub-alphabet K with n symbols. Consider a string compounded of the symbols of L and the variables x_1, ..., x_j. Let I(X) be the set of all instances of X, i.e. all strings resulting from substituting strings in K (i.e. constants) for the variables of X. And let $I_0(X)$ be the set of dyadic Gödel numbers of the strings in I(X).

Problem 2 [Substitution Lemma]. Prove that for each expression X compounded from the symbols of L, the set $I_0(X)$ is arithmetic.

We recall that by a *proof* in an elementary formal system (E) is meant a finite sequence X_1, ..., X_t of sentences of (E) such that each member X_i of the sequence is either an axiom (i.e. an instance of an initial formula of (E)), or is derivable from two earlier members of the sequence by what we will now call the rule of detachment: Y can be inferred from X and X → Y, providing that X is atomic.

We now assign to every finite sequence X_1, ..., X_t of expressions of L a number which we will call the *sequence number* of the sequence as follows. Let r be one larger than the number of symbols in L and let g* be a dyadic numeral starting with a 1 followed by a string of r 2's. Thus the string of 2's in g* will be longer than any string of 2's in the Gödel number of any expression compounded out of the alphabet of L, because those are assigned by listing the alphabet of L in some order, and then assigning consecutive dyadic numerals (starting with 1) as Gödel numbers to the symbols. Then we take the sequence number of X_1, ..., X_t to be g*g(X_1)g*g(X_2)g*... g*g(X_n)g*, where

$g(X_1)$, ..., $g(X_n)$ are the respective dyadic Gödel numbers of X_1, ..., X_t.

We let $seq(x)$ be the condition that x is a sequence number of some sequence X_1, ..., X_t. We let $x \in y$ be the condition that y is a sequence number of some sequence and x is the dyadic Gödel number of one of the terms of the sequence. We let $E(x, y, z)$ be the condition that z is the sequence number of some sequence z of which y is the Gödel number of some term Y of Z and x is the Gödel number of some term X which occurs earlier than the first occurrence of Y.

Problem 3. Prove the following to be arithmetic:

(a) $seq(y)$

(b) $x \in y$

(c) $E(x, y, z)$

We now define $Pf(x)$ to mean that x is the sequence number of a proof in (E), and we define $yPfx$ to mean that y is the sequence number of a proof X_1, ..., X_t and x is the Gödel number of X_t. In everyday mathematical terms, this means that y is the sequence number of a proof of X_t.

Let I_0 be the set of Gödel numbers of the initial sentences (the axioms) of (E) and let $Der(x, y, z)$ be the relation "x, y, z are the respective Gödel numbers of the expression X, Y, Z, where Z is directly derivable from X and Y by the rule of detachment—i.e. $Y = X \rightarrow Z$ and '$\rightarrow$' is not part of X."

Problem 4. Prove the following sets and relations are arithmetic.

(a) I_0

(b) The relation $Der(x, y, z)$

(c) $Pf(x)$ (x is the sequence number of a proof)

(d) y pf x (y is the sequence number of a proof of x—i.e. $Pf(y)$ and x is the Gödel number of the last term of the proof).

Problem 5. Prove that if W is a set representable in (E) then the set W_0 of dyadic Gödel numbers of the elements of W is arithmetic.

We now see by Problem 5 that the set T of true sentences of first-order arithmetic is not representable in any elementary formal system, for if it were, then the set T_0 of dyadic Gödel numbers of the true sentences would be arithmetic, which it isn't by Tarski's theorem proved in the last chapter.

Solutions to the Problems of Chapter XIX

1. $G_n(x)$ iff $\sim g_{n+1}Px \wedge 1Bx \wedge 2Ex \wedge \sim 11Px$

2. Recall that n is the number of symbols of K. Before proving the general case, let us first consider an example. Suppose X is the string $x_2a_1a_2x_1x_2a_1$. Here a_1 and a_2 are individual symbols of L that are not variables (or parts of variables) in X. We write their Gödel numbers as $g(a_1)$ and $g(a_2)$. Then

$x \in I_0$ iff $\exists x_1 \exists x_2 (G_n(x_1) \wedge G_n(x_2) \wedge$
$x_2 g(a_1)g(a_2)x_1x_2 g(a_1) = x)$.

More generally, Let X^* be the string that results when each symbol a_i of L (that is not a variable of X or part of a variable of X) is replaced in X by its Gödel number $g(a_i)$. Then if x_1, ..., x_j are the variables in X (which are still variables in X^*) we have

$x \in I_0$ iff $\exists x_1 ... \exists x_j (G_n(x_1) \wedge ... \wedge G_n(x_n) \wedge x = X^*)$.

3. (a) Seq(x) iff $g^*Bx \wedge g^*Ex \wedge \sim g^*g^*Px \wedge G_n(x)$

   (b) $x \in y$ iff Seq(y) $\wedge g^*xg^*Py \wedge G_n(x)$

   (c) E(x,y,z) iff Seq(z) $\wedge \exists w(wBz \wedge x \in w \wedge \sim y \in w)$

4. (a) Let Y_1, ..., Y_n be the initial formulas (axioms) of (E) and for each $i \leq n$, let A_i be the set of Gödel numbers of all the instances of Y_i. By the substitution lemma (Problem 2) each of the sets A_i is arithmetic. Let I_0 be the union $A_1 \cup ... \cup A_n$ of the sets A_1, ..., A_n—i.e. the set of all numbers x such that x belongs to at least one of the sets A_1, ..., A_n. This set I_0 is arithmetic, since $x \in I_0(X)$ iff $x \in A_1 \vee \cdots \vee x \in A_n$.

   (b) Let a be the Gödel number of the arrow, $\rightarrow$. Then Der(x, y, z) iff $y = xaz \wedge \sim aPx$.

   (c) Pf(w) iff

   Seq(w) $\wedge \forall z(z \in w \supset (I_0(z) \vee$ (continued on next line)
   $\exists x \exists y (E(x,z,w) \wedge E(y,z,w) \wedge Der(x,y,z)))$

   (d) ypfx iff Pf(y) $\wedge x \in y \wedge g^*xg^*Ey$

5. Let H be a predicate of (E) that represents the set W and let h be its Gödel number. Then $x \in W_0$ iff $\exists y(y \text{ pf } hx)$.

CHAPTER XX
THE INCOMPLETENESS OF PEANO ARITHMETIC

In 1891, Guiseppe Peano published his famous postulates for the positive integers (then sometimes called the "natural numbers," but nowadays the so-called "natural numbers" include zero) [Peano, 1891]. Here are the postulates (or axioms).

His undefined notions were 1 and the successor operation S(n).

1. 1 is a natural number.

2. If n is a natural number, so is S(n).

3. If S(n) = S(m) then n = m (no two distinct natural numbers have the same successor).

4. S(n) ≠ 1 (1 is not the successor of any natural number).

5. [Axiom of Mathematical Induction]. Suppose K is a set such that:

(a) 1 ∈ K.

(b) For any natural number n ∈ K, the natural number S(n) is also in K.

Then K contains all the natural numbers.

Peano's so-called "axioms" do not constitute an axiom system in the modern sense of the term. They might aptly be called "informal axioms," which, like the axioms of the ancient Greeks found in Euclid's mathematics books, consisted of truths regarded as self-evident.

Now we turn to the modern version of Peano Arithmetic in which the underlying logic is made explicit, and to which axioms for addition and multiplication are added to the axioms for succession.

The axioms fall into three groups—those of Group I are the axioms of *propositional* logic, which deal with the logical connectives ~, ∧, ∨ and ⊃ (not, and, or, implies). Those of Group II are axioms of what is called *first-order logic*, and are concerned with the quantifiers ∀ and ∃ ("for all" and "there exists"). The axioms of Group III are concerned with purely arithmetic notions—successor, plus and times.

Re Group I, there are many, many different axiom systems for proportional logic. They are all equivalent, in that the class of provable formulas are the same in each system. Some systems take all four connective ~, ∧, ∨ and ⊃ as undefined; others take some of them as undefined and define the others in terms of the undefined ones. For the sake of economy, we shall take ~ (negation) and ⊃ (implication) as undefined and for any formulas X and Y, we shall define X ∨ Y to be ~X ⊃ Y, and X ∧ Y to be ~(X ⊃ ~Y).

In general, a formal axiom system consists of a set of formulas called *axioms* together with a set of rules for deriving other formulas from the axioms—these rules are called *rules of inference*.

Now here is the axiom system for Peano Arithmetic that we will take. The axioms will be infinite in number, but each axiom will be of one of twelve easily recognizable forms called *axiom schemes*. In displaying these schemes, F, G and H are any formulas, x and y are any variables, and t is any term. For example, the first scheme L_1 means that for *any* pair of formulas F and G, the formula (F ⊃ (G ⊃ F)) is to be taken as an axiom; axiom scheme L_4

means that for any variable x and any pair of formulas F and G, the formula $(\forall x(F \supset G) \supset (\forall xF \supset \forall xG))$ is to be taken as an axiom.

Group I. Axiom Schemes for Propositional Logic

L_1: $(F \supset (G \supset F))$

L_2: $(F \supset (G \supset H)) \supset ((F \supset G) \supset (F \supset H)))$

L_3: $((\sim F \supset \sim G) \supset (G \supset F))$

Group II. Additional Axiom Schemes for First-Order Logic with Identity

L_4: $(\forall x(F \supset G) \supset (\forall xF \supset \forall xG))$

L_5: $(F \supset \forall xF)$, providing x does not occur in F.

L_6: $\exists x(x = t)$, providing x does not occur in t.

L_7: $((x = t) \supset (XxY \supset XtY))$, where X and Y are any expressions such that XxY is an *atomic* formula.

L_8: $(\forall xF \supset \forall x((x = t) \supset F))$

L_9: $(\sim \forall xF \supset \exists x \sim F)$

Group III. Six Axiom Schemes Having Only One Axiom Apiece, plus the Axiom Scheme of Mathematical Induction.

N_1: $((x' = y') \supset (x = y))$

N_2: $\sim(0 = x')$

N_3: $((x + 0) = x)$

N_4: $((x + y') = (x + y)')$

N_5: $((x \times 0) = 0)$

N_6: $((x \times y') = ((x \times y) + x))$

The next scheme consists of one axiom scheme—the scheme of Mathematical Induction—but infinitely many axioms—one for each formula F(x). Peano stated the axiom of mathematical induction as follows: For any formula F with a free variable x, if F(0) is true, and if it is also true for any number x, that F(x) being true implies F(x') is true, then F(x) is true for all natural numbers. We might try to write that out in our system as follows (dropping the outermost pair of parentheses for easier readability):

$(F(0) \land \forall x(F(x) \supset F(x'))) \supset \forall x F(x)$.

But since our version of Peano Arithmetic includes only the connectives $\sim$ and $\supset$, we can use the equivalence between $S_1 \land S_2 \supset S3$ and $(S_1 \supset S_2) \supset S_3$ that we saw in Chapter XVIII to rewrite the axiom of mathematical induction for our system as follows (again dropping outermost parentheses):

$F(0) \supset \forall x(F(x) \supset F(x')) \supset \forall x F(x)$.

Here $F(0)$ is the result of substituting the symbol 0 for every occurrence of x in F and $F(x')$ is the result of substituting the expression x' for every occurrence of x in F. But because we wish to show that the set of all provable formulas of Peano Arithmetic is representable in some elementary formal system (E), we are going to replace this axiom with one equivalent to it, namely one which avoids our having to represent the concept of substitution. We are going to use the same trick employed in Chapter XVIII during our work on arithmetizing the relation $g(\bar{x}) = y$ during the proof of Tarski's Theorem. Thus we define two new expressions:

$F[0] : \forall x((x = 0) \supset F)$

$F[x'] : \forall y((y = x') \supset \forall x((x = y) \supset F))$

The reader should be able to verify that $F[0]$ is equivalent to $F(0)$ and $F[x']$ is equivalent to $F(x')$. And F is *exactly the same* as $F(x)$, because in the latter use of the parentheses we are not talking about any substitution, but just suggesting that the formula F may have a free variable x. (But it doesn't have to! Although Peano stated his axiom of mathematical induction for a formula F with a free variable x, the reader should verify that the axiom of induction is true—although fruitless—even if F does not have x as a free variable, indeed if the variable x doesn't even occur in F at all.)

Substituting these equivalent formulas into our original rewriting of Peano, we have as our axiom of mathematical induction the following:

(F[0] ∧ ∀x(F ⊃ F[x'])) ⊃ ∀xF

Again using the equivalence from Chapter XVIII recently mentioned, that becomes the following when we wish (for simplicity's sake, to shorten a later proof) to avoid the ∧ that is not strictly part of our current version of Peano Arithmetic (now replacing the outermost pair of parentheses for greater precision):

N₇: ((F[0] ⊃ ∀x(F ⊃ F[x'])) ⊃ ∀xF).

The Inference Rules in our system of Peano Arithmetic are two in number:

Rule I [Rule of Detachment, also known as *Modus Ponens*.] From F and F ⊃ G to infer G.

Rule II [Generalization] From F to infer ∀xF.

A formula of Peano Arithmetic is called *provable* (in Peano Arithmetic) if its being so is a consequence of the following three conditions:

1. Every axiom is provable.
2. If F and F ⊃ G are provable, so is G.
3. If F is provable, so is ∀xF.

More explicitly, by a *proof* in P. A. (Peano Arithmetic) is meant a finite sequence of formulas such that each member of the sequence is either an axiom, or is directly derivable from two earlier members of the sequence by Rule I, or is directly derivable from one earlier member of the sequence by Rule II. A formula is then called *provable* if it is a member of the sequence of formulas of some proof.

We now wish to see that the set of provable formulas of Peano Arithmetic is representable in some elementary formal system.

Problem. Construct an elementary formal system (E) in which the set of provable formulas of Peano Arithmetic is represented. Do this in stages by successively showing the following items to be representable.

1. The set of strings of accents
2. The set of variables of P. A.
3. The relation "x and y are distinct variables"
4. The set of numerals
5. The set of terms
6. The set of atomic formulas
7. The set of formulas
8. The relation "x is a variable, t is a term and x does not occur in t"
9. The relation "x is a variable, f is a formula and x does not occur in f"
10. The set of axioms
11. The set of provable formulas of P. A.

* * *

Having constructed an elementary formal system in which the set P of provable formulas of Peano Arithmetic is represented, we go about getting an undecidable sentence of Peano Arithmetic as follows:

By the last chapter, we know how to get an arithmetic formula that expresses the set P_0 of dyadic Gödel numbers of all the elements of P. Its negation then expresses the complement $\overline{P_0}$ of P_0. Then we know how to get a formula $F(v_1)$ that expresses the set $\widetilde{P_0}^\#$ (Problem 11, Chapter XVIII). This formula has a g-number h, and the sentence $F[\overline{h}]$ (square brackets) is then a Gödel sentence for the set $\widetilde{P_0}$. Then, as seen in the solution of Problem 15, Chapter XVIII, the sentence $F[\overline{h}]$ is true if and only if $F[\overline{h}]$ is not in the set P, which means that $F[\overline{h}]$ is either true and not provable (not in P), or false but provable. Under the reasonable assumptions that no false sentences are provable, the sentence F[h̄] is thus true but not provable in Peano Arithmetic. Since the negation ~F[h̄]] of the true sentence F[h̄] is false, it is also not provable in Peano Arithmetic, and is thus an undecidable sentence of Peano Arithmetic.

The proof that we have given of the incompleteness of Peano Arithmetic is based on the assumption that all sentences provable in Peano Arithmetic are true. How do we know that? Well, here is an argument to support this: Given a formula F, by an *instance* of F is meant any sentence resulting from substituting numbers for all free occurrences of all the variables of F. If the formula is a sentence (i.e. has no free variables) then its only instance is itself. In the last chapter we defined what it means for a sentence to be *true*. Let us now call a formula F *correct* if all its instances are true. Now, it is obvious from inspection that all the axioms of Peano Arithmetic are correct and that the inference rules preserve correctness—i.e. that if F and F ⊃ G are both correct, so is G, and if F is correct, so is ∀xF. It then follows by mathematical induction that every provable formula is correct, hence every provable sentence is true.

As remarked earlier, Gödel's original proof did not involve the notion of truth, but rather of the of the weaker notion of omega consistency. Then Rosser's proof involves only the still weaker notion of simple consistency. We shall not take the space here to give the original Gödel proof, or the Rosser proof for the system of Peano Arithmetic. The reader can find all details of these proofs in my book "Gödel's Incompleteness Theorems," as well as many other books on mathematical logic.

In the proof I have given here that the set of Gödel numbers of the provable formulas is arithmetic, I have used elementary formal systems. There are many other ways of obtaining an arithmetic formula for P_0 that do not use elementary formal systems. A particularly direct and simple way is in my book "Gödel's Incompleteness Theorems" [Smullyan, 1992] which uses ideas due to Quine.

* * *

In conclusion, I wish to express some personal opinions: To me, the incompleteness of Peano Arithmetic is less interesting than the fact that the set of true sentences of first-order arithmetic is not recursively enumerable! Most mathematicians are interested in which arithmetic sentences are *true,* and which are not. No number theorist ever works with any formal axiom system such as Peano Arithmetic. If a sentence turns out to be provable in Peano Arithmetic, that's all to the good, since we then know that the sentence is true. But what about the myriad of true sentences that are not provable in Peano Arithmetic? It is those sentences that are of interest to working mathematicians.

The important thing about the fact that the set of true sentences of first-order arithmetic is not even recursively enumerable, let alone solvable, means that no possible purely mechanical device can tell us which sentences are true and which are not. This means that ingenuity is required for the discovery of mathematical truths. In the prophetic words of the logician Emil Post, "this means that we should perhaps go back to the nineteenth century attitude that truth and meaning is the essence of mathematics. Mathematics is and must remain an essentially creative endeavor."

Solution to the Problem of Chapter XX

In what follows, $\supset$ is the implication symbol of Peano Arithmetic, whereas $\rightarrow$ is the implication symbol of the elementary formal system (E) that we are constructing. We use letters x, y, z, w, t, f, g, sometimes with subscripts as variables of the elementary formal system (E), as opposed to the variables (v'), (v''), etc. of Peano Arithmetic. The alphabet K of (E) will be the set of all the symbols used to create formulas in first-order arithmetic.

We introduce the predicates and initial formulas (axioms of (E)) in groups, first explaining what each newly introduced predicate is to represent.

acc represents the set of strings of accents

acc '

acc x → acc x'

V represents the set of variables

acc x → V(vx)

D represents the relation "x and y are distinct variables"

acc x → acc y → D(vx), (vxy)

Dx, y → Dy, x

N represents the set of numerals.

N0

Acc x → Ny → Nyx

T represents the set of terms

Vx → Tx

Nx → Tx

Tx → Tx'

Tx → Ty → Tx + y

Tx → Ty → Tx × y

F_0 represents the set of atomic formulas

Tx → Ty → F_0(x = y)

F represents the set of formulas

F_0x → Fx

$Fx \to F{\sim}x$

$Fx \to Fy \to F(x \supset y)$

$Fx \to Vy \to F\forall yx$

$Fx \to Vy \to F\exists yx$

Q_0 represents the relation "x is a variable, t is a term and x does not occur in t"

$Vx \to Ny \to Q_0x,y$

$Dx,y \to Q_0x,y$

$Q_0x,t \to Q_0x,t'$

$Q_0x,t_1 \to Q_0x,t_2 \to Q_0x,t_1 + t_2$

$Q_0x,t_1 \to Q_0x,t_2 \to Q_0x,t_1 \times t_2$

Q represents the relation: "x is a variable, f is a formula and x does not occur in f."

$Q_0x,y \to Q_0x,z \to Qx,y = z$

$Qx,f \to Qx,{\sim}f$

$Qx,f \to Qx,g \to Qx,f \supset g$

$Qx,f \to Dx,y \to Qx,\forall yf$

$Qx,f \to Dx,y \to Qx,\exists yf$

A represents the set of axioms of P.A.

$Fx \to Fy \to A(x \supset (y \supset x)\}$

$Fx \to Fy \to Fz \to A((x \supset (y \supset z)) \supset ((x \supset y) \supset (x \supset z)))$

$Fx \to Fy \to A((\sim x \supset \sim y) \supset (y \supset x))$

$Vx \to Ff \to Fg \to A(\forall x(f \supset g) \supset (\forall xf \supset \forall xg))$

$Qx,f \to A(f \supset \forall xf)$

$Q_0x,t \to A\exists x(x = t)$

$Vx \to Tt \to F_0yxz \to A((x = t) \supset (yxz \supset ytz))$

$Vx \to Tt \to Ff \to A(\forall xf \supset \forall x((x = t) \supset f))$

$Vx \to Ff \to A(\sim\forall xf \supset \exists x\sim fx)$

$Vx \to Vy \to A((x' = y') \supset (x = y))$

$Vx \to A{\sim}(0 = x')$

$Vx \to A((x + 0) = x)$

$Vx \to Vy \to A((x + y') = (x + y)')$

$Vx \to A((x \times 0) = 0)$

Vx → Vy → A((x × y') = ((x × y) + x))

In the next axiom of (E), I will use the following abbreviations:

f[0]   for $\forall x((x = 0) \supset f)$
f '   for $\forall y((y = x' \supset \forall x((x = y) \supset f)$

Thus the axiom of mathematical induction can be represented by the following formula:

Dx,y → Qy,f → A((f[0] ⊃ ∀x(f ⊃ f')) ⊃ ∀xf)

P represents the set of provable formulas.

Af → Pf

Pf → Pf ⊃ g → Pg

Pf → Vx → P∀xF

REFERENCES

Gödel, Kurt, 1931. Über formal unentsheidbarc Sätze der "Principia Mathematica" und verwandter Systeme, *Monatshefte für Mathematik und Physik* 38: pp. 173–198.

Peano, Giuseppe, 1891. Sul concetto di numero, *Rivista di Matematica, Vol.* 1, pp. 87–102.

Quine, W. V., 1946. Concatenation as a Basis for Arithmetic, *The Journal of Symbolic Logic, Vol.* 11, # 4, pp. 105–114.

Rosser, J. Barkley, 1936. Extensions of some Theorems of Gödel and Church, *Journal of Symbolic Logic, Vol.* 1, pp. 87–91.

Smullyan, Raymond, 1961. *Theory of Formal Systems,* Princeton University Press.

Smullyan, Raymond, 1992. *Gödel's Incompleteness Theorems,* Oxford University Press.

Tarski, Alfred, 1932. Der Wahrsheitsbegriff in den formalen Sprachen der deductiven Disziplinen. *Studia Philosophica, Vol.* 1, pp. 261–405.

Tarski, Alfred, 1953. *Undecidable Theories,* North Holland Publishing Company.

A CATALOG OF SELECTED
DOVER BOOKS
IN SCIENCE AND MATHEMATICS

Mathematics-Bestsellers

HANDBOOK OF MATHEMATICAL FUNCTIONS: with Formulas, Graphs, and Mathematical Tables, Edited by Milton Abramowitz and Irene A. Stegun. A classic resource for working with special functions, standard trig, and exponential logarithmic definitions and extensions, it features 29 sets of tables, some to as high as 20 places. 1046pp. 8 x 10 1/2.   0-486-61272-4

ABSTRACT AND CONCRETE CATEGORIES: The Joy of Cats, Jiri Adamek, Horst Herrlich, and George E. Strecker. This up-to-date introductory treatment employs category theory to explore the theory of structures. Its unique approach stresses concrete categories and presents a systematic view of factorization structures. Numerous examples. 1990 edition, updated 2004. 528pp. 6 1/8 x 9 1/4.   0-486-46934-4

MATHEMATICS: Its Content, Methods and Meaning, A. D. Aleksandrov, A. N. Kolmogorov, and M. A. Lavrent'ev. Major survey offers comprehensive, coherent discussions of analytic geometry, algebra, differential equations, calculus of variations, functions of a complex variable, prime numbers, linear and non-Euclidean geometry, topology, functional analysis, more. 1963 edition. 1120pp. 5 3/8 x 8 1/2.   0-486-40916-3

INTRODUCTION TO VECTORS AND TENSORS: Second Edition-Two Volumes Bound as One, Ray M. Bowen and C.-C. Wang. Convenient single-volume compilation of two texts offers both introduction and in-depth survey. Geared toward engineering and science students rather than mathematicians, it focuses on physics and engineering applications. 1976 edition. 560pp. 6 1/2 x 9 1/4.   0-486-46914-X

AN INTRODUCTION TO ORTHOGONAL POLYNOMIALS, Theodore S. Chihara. Concise introduction covers general elementary theory, including the representation theorem and distribution functions, continued fractions and chain sequences, the recurrence formula, special functions, and some specific systems. 1978 edition. 272pp. 5 3/8 x 8 1/2.
                                                                            0-486-47929-3

ADVANCED MATHEMATICS FOR ENGINEERS AND SCIENTISTS, Paul DuChateau. This primary text and supplemental reference focuses on linear algebra, calculus, and ordinary differential equations. Additional topics include partial differential equations and approximation methods. Includes solved problems. 1992 edition. 400pp. 7 1/2 x 9 1/4.   0-486-47930-7

PARTIAL DIFFERENTIAL EQUATIONS FOR SCIENTISTS AND ENGINEERS, Stanley J. Farlow. Practical text shows how to formulate and solve partial differential equations. Coverage of diffusion-type problems, hyperbolic-type problems, elliptic-type problems, numerical and approximate methods. Solution guide available upon request. 1982 edition. 414pp. 6 1/8 x 9 1/4.   0-486-67620-X

VARIATIONAL PRINCIPLES AND FREE-BOUNDARY PROBLEMS, Avner Friedman. Advanced graduate-level text examines variational methods in partial differential equations and illustrates their applications to free-boundary problems. Features detailed statements of standard theory of elliptic and parabolic operators. 1982 edition. 720pp. 6 1/8 x 9 1/4.   0-486-47853-X

LINEAR ANALYSIS AND REPRESENTATION THEORY, Steven A. Gaal. Unified treatment covers topics from the theory of operators and operator algebras on Hilbert spaces; integration and representation theory for topological groups; and the theory of Lie algebras, Lie groups, and transform groups. 1973 edition. 704pp. 6 1/8 x 9 1/4.
                                                                            0-486-47851-3

A SURVEY OF INDUSTRIAL MATHEMATICS, Charles R. MacCluer. Students learn how to solve problems they'll encounter in their professional lives with this concise single-volume treatment. It employs MATLAB and other strategies to explore typical industrial problems. 2000 edition. 384pp. 5 3/8 x 8 1/2. 0-486-47702-9

NUMBER SYSTEMS AND THE FOUNDATIONS OF ANALYSIS, Elliott Mendelson. Geared toward undergraduate and beginning graduate students, this study explores natural numbers, integers, rational numbers, real numbers, and complex numbers. Numerous exercises and appendixes supplement the text. 1973 edition. 368pp. 5 3/8 x 8 1/2. 0-486-45792-3

A FIRST LOOK AT NUMERICAL FUNCTIONAL ANALYSIS, W. W. Sawyer. Text by renowned educator shows how problems in numerical analysis lead to concepts of functional analysis. Topics include Banach and Hilbert spaces, contraction mappings, convergence, differentiation and integration, and Euclidean space. 1978 edition. 208pp. 5 3/8 x 8 1/2. 0-486-47882-3

FRACTALS, CHAOS, POWER LAWS: Minutes from an Infinite Paradise, Manfred Schroeder. A fascinating exploration of the connections between chaos theory, physics, biology, and mathematics, this book abounds in award-winning computer graphics, optical illusions, and games that clarify memorable insights into self-similarity. 1992 edition. 448pp. 6 1/8 x 9 1/4. 0-486-47204-3

SET THEORY AND THE CONTINUUM PROBLEM, Raymond M. Smullyan and Melvin Fitting. A lucid, elegant, and complete survey of set theory, this three-part treatment explores axiomatic set theory, the consistency of the continuum hypothesis, and forcing and independence results. 1996 edition. 336pp. 6 x 9. 0-486-47484-4

DYNAMICAL SYSTEMS, Shlomo Sternberg. A pioneer in the field of dynamical systems discusses one-dimensional dynamics, differential equations, random walks, iterated function systems, symbolic dynamics, and Markov chains. Supplementary materials include PowerPoint slides and MATLAB exercises. 2010 edition. 272pp. 6 1/8 x 9 1/4. 0-486-47705-3

ORDINARY DIFFERENTIAL EQUATIONS, Morris Tenenbaum and Harry Pollard. Skillfully organized introductory text examines origin of differential equations, then defines basic terms and outlines general solution of a differential equation. Explores integrating factors; dilution and accretion problems; Laplace Transforms; Newton's Interpolation Formulas, more. 818pp. 5 3/8 x 8 1/2. 0-486-64940-7

MATROID THEORY, D. J. A. Welsh. Text by a noted expert describes standard examples and investigation results, using elementary proofs to develop basic matroid properties before advancing to a more sophisticated treatment. Includes numerous exercises. 1976 edition. 448pp. 5 3/8 x 8 1/2. 0-486-47439-9

THE CONCEPT OF A RIEMANN SURFACE, Hermann Weyl. This classic on the general history of functions combines function theory and geometry, forming the basis of the modern approach to analysis, geometry, and topology. 1955 edition. 208pp. 5 3/8 x 8 1/2. 0-486-47004-0

THE LAPLACE TRANSFORM, David Vernon Widder. This volume focuses on the Laplace and Stieltjes transforms, offering a highly theoretical treatment. Topics include fundamental formulas, the moment problem, monotonic functions, and Tauberian theorems. 1941 edition. 416pp. 5 3/8 x 8 1/2. 0-486-47755-X

Browse over 9,000 books at www.doverpublications.com

Mathematics–Logic and Problem Solving

PERPLEXING PUZZLES AND TANTALIZING TEASERS, Martin Gardner. Ninety-three riddles, mazes, illusions, tricky questions, word and picture puzzles, and other challenges offer hours of entertainment for youngsters. Filled with rib-tickling drawings. Solutions. 224pp. 5 3/8 x 8 1/2.　　　　　　　0-486-25637-5

MY BEST MATHEMATICAL AND LOGIC PUZZLES, Martin Gardner. The noted expert selects 70 of his favorite "short" puzzles. Includes The Returning Explorer, The Mutilated Chessboard, Scrambled Box Tops, and dozens more. Complete solutions included. 96pp. 5 3/8 x 8 1/2.　　　　　　　0-486-28152-3

THE LADY OR THE TIGER?: and Other Logic Puzzles, Raymond M. Smullyan. Created by a renowned puzzle master, these whimsically themed challenges involve paradoxes about probability, time, and change; metapuzzles; and self-referentiality. Nineteen chapters advance in difficulty from relatively simple to highly complex. 1982 edition. 240pp. 5 3/8 x 8 1/2.　　　　　　　0-486-47027-X

SATAN, CANTOR AND INFINITY: Mind-Boggling Puzzles, Raymond M. Smullyan. A renowned mathematician tells stories of knights and knaves in an entertaining look at the logical precepts behind infinity, probability, time, and change. Requires a strong background in mathematics. Complete solutions. 288pp. 5 3/8 x 8 1/2.

0-486-47036-9

THE RED BOOK OF MATHEMATICAL PROBLEMS, Kenneth S. Williams and Kenneth Hardy. Handy compilation of 100 practice problems, hints and solutions indispensable for students preparing for the William Lowell Putnam and other mathematical competitions. Preface to the First Edition. Sources. 1988 edition. 192pp. 5 3/8 x 8 1/2.　　　　　　　0-486-69415-1

KING ARTHUR IN SEARCH OF HIS DOG AND OTHER CURIOUS PUZZLES, Raymond M. Smullyan. This fanciful, original collection for readers of all ages features arithmetic puzzles, logic problems related to crime detection, and logic and arithmetic puzzles involving King Arthur and his Dogs of the Round Table. 160pp. 5 3/8 x 8 1/2.

0-486-47435-6

UNDECIDABLE THEORIES: Studies in Logic and the Foundation of Mathematics, Alfred Tarski in collaboration with Andrzej Mostowski and Raphael M. Robinson. This well-known book by the famed logician consists of three treatises: "A General Method in Proofs of Undecidability," "Undecidability and Essential Undecidability in Mathematics," and "Undecidability of the Elementary Theory of Groups." 1953 edition. 112pp. 5 3/8 x 8 1/2.　　　　　　　0-486-47703-7

LOGIC FOR MATHEMATICIANS, J. Barkley Rosser. Examination of essential topics and theorems assumes no background in logic. "Undoubtedly a major addition to the literature of mathematical logic." – *Bulletin of the American Mathematical Society*. 1978 edition. 592pp. 6 1/8 x 9 1/4.　　　　　　　0-486-46898-4

INTRODUCTION TO PROOF IN ABSTRACT MATHEMATICS, Andrew Wohlgemuth. This undergraduate text teaches students what constitutes an acceptable proof, and it develops their ability to do proofs of routine problems as well as those requiring creative insights. 1990 edition. 384pp. 6 1/2 x 9 1/4.　　　0-486-47854-8

FIRST COURSE IN MATHEMATICAL LOGIC, Patrick Suppes and Shirley Hill. Rigorous introduction is simple enough in presentation and context for wide range of students. Symbolizing sentences; logical inference; truth and validity; truth tables; terms, predicates, universal quantifiers; universal specification and laws of identity; more. 288pp. 5 3/8 x 8 1/2.　　　　　　　0-486-42259-3

Browse over 9,000 books at www.doverpublications.com

Mathematics–Algebra and Calculus

VECTOR CALCULUS, Peter Baxandall and Hans Liebeck. This introductory text offers a rigorous, comprehensive treatment. Classical theorems of vector calculus are amply illustrated with figures, worked examples, physical applications, and exercises with hints and answers. 1986 edition. 560pp. 5 3/8 x 8 1/2. 0-486-46620-5

ADVANCED CALCULUS: An Introduction to Classical Analysis, Louis Brand. A course in analysis that focuses on the functions of a real variable, this text introduces the basic concepts in their simplest setting and illustrates its teachings with numerous examples, theorems, and proofs. 1955 edition. 592pp. 5 3/8 x 8 1/2. 0-486-44548-8

ADVANCED CALCULUS, Avner Friedman. Intended for students who have already completed a one-year course in elementary calculus, this two-part treatment advances from functions of one variable to those of several variables. Solutions. 1971 edition. 432pp. 5 3/8 x 8 1/2. 0-486-45795-8

METHODS OF MATHEMATICS APPLIED TO CALCULUS, PROBABILITY, AND STATISTICS, Richard W. Hamming. This 4-part treatment begins with algebra and analytic geometry and proceeds to an exploration of the calculus of algebraic functions and transcendental functions and applications. 1985 edition. Includes 310 figures and 18 tables. 880pp. 6 1/2 x 9 1/4. 0-486-43945-3

BASIC ALGEBRA I: Second Edition, Nathan Jacobson. A classic text and standard reference for a generation, this volume covers all undergraduate algebra topics, including groups, rings, modules, Galois theory, polynomials, linear algebra, and associative algebra. 1985 edition. 528pp. 6 1/8 x 9 1/4. 0-486-47189-6

BASIC ALGEBRA II: Second Edition, Nathan Jacobson. This classic text and standard reference comprises all subjects of a first-year graduate-level course, including in-depth coverage of groups and polynomials and extensive use of categories and functors. 1989 edition. 704pp. 6 1/8 x 9 1/4. 0-486-47187-X

CALCULUS: An Intuitive and Physical Approach (Second Edition), Morris Kline. Application-oriented introduction relates the subject as closely as possible to science with explorations of the derivative; differentiation and integration of the powers of x; theorems on differentiation, antidifferentiation; the chain rule; trigonometric functions; more. Examples. 1967 edition. 960pp. 6 1/2 x 9 1/4. 0-486-40453-6

ABSTRACT ALGEBRA AND SOLUTION BY RADICALS, John E. Maxfield and Margaret W. Maxfield. Accessible advanced undergraduate-level text starts with groups, rings, fields, and polynomials and advances to Galois theory, radicals and roots of unity, and solution by radicals. Numerous examples, illustrations, exercises, appendixes. 1971 edition. 224pp. 6 1/8 x 9 1/4. 0-486-47723-1

AN INTRODUCTION TO THE THEORY OF LINEAR SPACES, Georgi E. Shilov. Translated by Richard A. Silverman. Introductory treatment offers a clear exposition of algebra, geometry, and analysis as parts of an integrated whole rather than separate subjects. Numerous examples illustrate many different fields, and problems include hints or answers. 1961 edition. 320pp. 5 3/8 x 8 1/2. 0-486-63070-6

LINEAR ALGEBRA, Georgi E. Shilov. Covers determinants, linear spaces, systems of linear equations, linear functions of a vector argument, coordinate transformations, the canonical form of the matrix of a linear operator, bilinear and quadratic forms, and more. 387pp. 5 3/8 x 8 1/2. 0-486-63518-X

Mathematics-Probability and Statistics

BASIC PROBABILITY THEORY, Robert B. Ash. This text emphasizes the probabilistic way of thinking, rather than measure-theoretic concepts. Geared toward advanced undergraduates and graduate students, it features solutions to some of the problems. 1970 edition. 352pp. 5 3/8 x 8 1/2. 0-486-46628-0

PRINCIPLES OF STATISTICS, M. G. Bulmer. Concise description of classical statistics, from basic dice probabilities to modern regression analysis. Equal stress on theory and applications. Moderate difficulty; only basic calculus required. Includes problems with answers. 252pp. 5 5/8 x 8 1/4. 0-486-63760-3

OUTLINE OF BASIC STATISTICS: Dictionary and Formulas, John E. Freund and Frank J. Williams. Handy guide includes a 70-page outline of essential statistical formulas covering grouped and ungrouped data, finite populations, probability, and more, plus over 1,000 clear, concise definitions of statistical terms. 1966 edition. 208pp. 5 3/8 x 8 1/2. 0-486-47769-X

GOOD THINKING: The Foundations of Probability and Its Applications, Irving J. Good. This in-depth treatment of probability theory by a famous British statistician explores Keynesian principles and surveys such topics as Bayesian rationality, corroboration, hypothesis testing, and mathematical tools for induction and simplicity. 1983 edition. 352pp. 5 3/8 x 8 1/2. 0-486-47438-0

INTRODUCTION TO PROBABILITY THEORY WITH CONTEMPORARY APPLICATIONS, Lester L. Helms. Extensive discussions and clear examples, written in plain language, expose students to the rules and methods of probability. Exercises foster problem-solving skills, and all problems feature step-by-step solutions. 1997 edition. 368pp. 6 1/2 x 9 1/4. 0-486-47418-6

CHANCE, LUCK, AND STATISTICS, Horace C. Levinson. In simple, non-technical language, this volume explores the fundamentals governing chance and applies them to sports, government, and business. "Clear and lively ... remarkably accurate." – *Scientific Monthly.* 384pp. 5 3/8 x 8 1/2. 0-486-41997-5

FIFTY CHALLENGING PROBLEMS IN PROBABILITY WITH SOLUTIONS, Frederick Mosteller. Remarkable puzzlers, graded in difficulty, illustrate elementary and advanced aspects of probability. These problems were selected for originality, general interest, or because they demonstrate valuable techniques. Also includes detailed solutions. 88pp. 5 3/8 x 8 1/2. 0-486-65355-2

EXPERIMENTAL STATISTICS, Mary Gibbons Natrella. A handbook for those seeking engineering information and quantitative data for designing, developing, constructing, and testing equipment. Covers the planning of experiments, the analyzing of extreme-value data; and more. 1966 edition. Index. Includes 52 figures and 76 tables. 560pp. 8 3/8 x 11. 0-486-43937-2

STOCHASTIC MODELING: Analysis and Simulation, Barry L. Nelson. Coherent introduction to techniques also offers a guide to the mathematical, numerical, and simulation tools of systems analysis. Includes formulation of models, analysis, and interpretation of results. 1995 edition. 336pp. 6 1/8 x 9 1/4. 0-486-47770-3

INTRODUCTION TO BIOSTATISTICS: Second Edition, Robert R. Sokal and F. James Rohlf. Suitable for undergraduates with a minimal background in mathematics, this introduction ranges from descriptive statistics to fundamental distributions and the testing of hypotheses. Includes numerous worked-out problems and examples. 1987 edition. 384pp. 6 1/8 x 9 1/4. 0-486-46961-1

Mathematics-Geometry and Topology

PROBLEMS AND SOLUTIONS IN EUCLIDEAN GEOMETRY, M. N. Aref and William Wernick. Based on classical principles, this book is intended for a second course in Euclidean geometry and can be used as a refresher. More than 200 problems include hints and solutions. 1968 edition. 272pp. 5 3/8 x 8 1/2.   0-486-47720-7

TOPOLOGY OF 3-MANIFOLDS AND RELATED TOPICS, Edited by M. K. Fort, Jr. With a New Introduction by Daniel Silver. Summaries and full reports from a 1961 conference discuss decompositions and subsets of 3-space; n-manifolds; knot theory; the Poincaré conjecture; and periodic maps and isotopies. Familiarity with algebraic topology required. 1962 edition. 272pp. 6 1/8 x 9 1/4.   0-486-47753-3

POINT SET TOPOLOGY, Steven A. Gaal. Suitable for a complete course in topology, this text also functions as a self-contained treatment for independent study. Additional enrichment materials make it equally valuable as a reference. 1964 edition. 336pp. 5 3/8 x 8 1/2.   0-486-47222-1

INVITATION TO GEOMETRY, Z. A. Melzak. Intended for students of many different backgrounds with only a modest knowledge of mathematics, this text features self-contained chapters that can be adapted to several types of geometry courses. 1983 edition. 240pp. 5 3/8 x 8 1/2.   0-486-46626-4

TOPOLOGY AND GEOMETRY FOR PHYSICISTS, Charles Nash and Siddhartha Sen. Written by physicists for physics students, this text assumes no detailed background in topology or geometry. Topics include differential forms, homotopy, homology, cohomology, fiber bundles, connection and covariant derivatives, and Morse theory. 1983 edition. 320pp. 5 3/8 x 8 1/2.   0-486-47852-1

BEYOND GEOMETRY: Classic Papers from Riemann to Einstein, Edited with an Introduction and Notes by Peter Pesic. This is the only English-language collection of these 8 accessible essays. They trace seminal ideas about the foundations of geometry that led to Einstein's general theory of relativity. 224pp. 6 1/8 x 9 1/4. 0-486-45350-2

GEOMETRY FROM EUCLID TO KNOTS, Saul Stahl. This text provides a historical perspective on plane geometry and covers non-neutral Euclidean geometry, circles and regular polygons, projective geometry, symmetries, inversions, informal topology, and more. Includes 1,000 practice problems. Solutions available. 2003 edition. 480pp. 6 1/8 x 9 1/4.   0-486-47459-3

TOPOLOGICAL VECTOR SPACES, DISTRIBUTIONS AND KERNELS, François Trèves. Extending beyond the boundaries of Hilbert and Banach space theory, this text focuses on key aspects of functional analysis, particularly in regard to solving partial differential equations. 1967 edition. 592pp. 5 3/8 x 8 1/2.
0-486-45352-9

INTRODUCTION TO PROJECTIVE GEOMETRY, C. R. Wylie, Jr. This introductory volume offers strong reinforcement for its teachings, with detailed examples and numerous theorems, proofs, and exercises, plus complete answers to all odd-numbered end-of-chapter problems. 1970 edition. 576pp. 6 1/8 x 9 1/4.   0-486-46895-X

FOUNDATIONS OF GEOMETRY, C. R. Wylie, Jr. Geared toward students preparing to teach high school mathematics, this text explores the principles of Euclidean and non-Euclidean geometry and covers both generalities and specifics of the axiomatic method. 1964 edition. 352pp. 6 x 9.   0-486-47214-0

Browse over 9,000 books at www.doverpublications.com